BOOK II.4

THE SINGULARITY DUAL-ASPECT EMERGENCE (SDAE) CANON

# THE MATHEMATICS OF BECOMING

*Meaning, Coherence, Wisdom, Dual Aspect, and Recursive Scale*

*PRINT VOLUME 4 OF 15*

ASTON GUSTAVOUS FARQUHARSON

NuSpecies Press

AF866037

The Mathematics of Becoming: Meaning, Coherence, Wisdom, Dual Aspect, and Recursive Scale

Copyright © 2026 Aston Gustavous Farquharson

All rights reserved.

Published by NuSpecies Press

Pawling, New York, USA

First Edition

eBook ISBN: 979-8-952413-00-9

Paperback ISBN: 979-8-952413-01-6

Hardback ISBN: 979-8-952413-02-3

## EDITION AND RIGHTS

*First public edition, 2026. Book II.4 is Volume 4 of the fifteen-volume public edition of The Mathematics of Becoming and forms part of The Singularity Dual-Aspect Emergence (SDAE) Canon. Each physical and digital format is issued under its own ISBN and format-specific metadata.*

*AI-assisted editorial, research-organization, design, and production tools supported this work under the author's direction. Aston Gustavous Farquharson established the intellectual architecture, selected and revised the content, reviewed the claims and sources, and approved the final text and figures.*

# CONTENTS

*BOOK II.4 - volume-local folios*



*Front matter is unnumbered. Controlled blank leaves carry no folio.*

# LIST OF FIGURES

*Permanent figure identifiers govern citation*



*All listed locations contain source-locked public figures. FS identifies figure position; EQ identifies canonical equation.*

# HOW TO READ THIS VOLUME

## ONE VOLUME WITHIN A CONTINUOUS ARCHITECTURE

*Book II.4 can be read independently, while its chapters, equations, figures, and permanent identifiers remain continuous with the wider Book II sequence.*

## EVIDENCE AND CLAIM LEVELS

*Established science, inherited mathematics, conventional models, SDAE formal constructions, architectural recognitions, lived-experience descriptions, ethical constraints, and symbolic teachings are intentionally distinguished. Their proximity does not make them evidentially identical.*

## COHERENCE, CONSCIOUSNESS, AND DUAL-ASPECT BOUNDARIES

*Meaning, coherence, wisdom, exterior and interior state spaces, recursive models, ratios, and balance conditions are domain-specific formal or symbolic tools. Coherence does not prove moral goodness or consciousness; dual aspect does not establish two substances; and structural resemblance across scales does not prove one mechanism.*

## CONTINUING HUMAN FOUNDATION

*The SDAE Canon is rooted in the spiritual inheritance of parents and grandparents and in decades of face-to-face work with thousands of NuSpecians and their families confronting illness, suffering, caregiving, recovery, dying, and grief. These lived realities give the Canon its questions, compassion, and moral urgency; they do not substitute for controlled empirical evidence or establish a physical mechanism. Some dimensions of human experience - including meaning, interiority, dignity, hope, grief, relational coherence, spiritual orientation, and compassionate presence - may not be fully captured by present measures. This volume therefore asks how meaning, coherence, wisdom, inner-outer description, recursion, and dynamic balance may be formalized without reducing the whole person to a score, model, or mechanism.*

*This volume continues the dedication made in Book II.1 to the author's mother, father, and grandparents, whose faith taught that spirituality must be lived through compassion, presence, and dignity within this lifetime.*

## EQUATIONS, FIGURES, AND FIGURE SLOTS

*EQ-CC.NN identifies a canonical equation. G-NN identifies a canonical figure. FS-GNN.SS identifies a permanent position within a figure. Supporting equations are mapped by function rather than by matching ordinal numbers.*

## FIGURES AND ACCESSIBILITY

*Public figures are designed for grayscale reproduction and are accompanied by structured alternative descriptions in the accessible edition.*

*REALITY REMAINS THE MEASURE.*

# ABOUT THIS VOLUME

*Chapters Twenty-One through Twenty-Five*

This volume asks how knowledge becomes orientation, how organized identity persists through generation and loss, how judgment acts under uncertainty, how one reality may be approached through exterior and interior state spaces, and how patterns recur across time and scale without requiring one universal mechanism.

Meaning does not become a physical force; coherence does not become perfection; wisdom does not become optimization; dual aspect does not create two universes; and balance does not require a fifty-fifty ratio.

*Reality remains the measure.*

# BOOK II - FIFTEEN-VOLUME PUBLIC ARCHITECTURE

*II.1 Foundations of Mathematical Becoming*

*II.2 Reality, Relation, Differentiation, Boundary, and Vulnerability*

*II.3 Adaptation, Attention, Awareness, Learning, Memory, and Knowledge*

*II.4 Meaning, Coherence, Wisdom, Dual Aspect, and Recursive Scale - THIS VOLUME*

*II.5 Distributed Encoding, Bridge Species Responsibility, and Experimental Accountability*

*II.6 Complete Equation Registry I*

*II.7 Complete Equation Registry II and Master Crosswalk*

*II.8 Figure Production Constitution and Foundational Claim Architectures*

*II.9 Primordial Possibility, Fundamental Laws, Coherence Scaffolds, and Ma'at*

*II.10 Pre-Geometric Relation, Geometrogenesis, and Differentiated Identity*

*II.11 Boundary, Vulnerability, and Adaptive Recovery*

*II.12 Attention, Awareness, and Learning*

*II.13 Memory, Knowledge, and Meaning*

*II.14 Coherence, Judgment, and Dual-Aspect State Spaces*

*II.15 Recursion, Distributed Encoding, Bridge Species, and Computational Experiments*

*The complete archival master remains preserved as the intellectual and production concordance. Public volumes are curated for printable length, local navigation, and stable citation.*

# CANONICAL CITATION AND IDENTIFIER POLICY

Each public volume restarts physical folios at page 1. Stable intellectual identifiers remain primary: chapter number, permanent equation ID (EQ-CC.NN), figure ID (G-NN), figure-slot ID (FS-GNN.SS), and appendix or registry section.

Recommended citation form: Farquharson, The Mathematics of Becoming, Book II.4, Chapter 22, G-19; add the local page number shown in the edition consulted.

Cross-volume references must not rely upon page number alone. A later pagination correction may change a local folio, but it must never change a permanent equation, figure, or registry identifier.

## CLAIM-STATUS, CONSCIOUSNESS, AND HUMAN-DIGNITY DISCIPLINE

*Formal notation in this volume may represent established mathematics, a conventional domain model, an SDAE formal model, an architectural recognition, an ethical constraint, or a symbolic teaching. The surrounding label and source record determine its status; notation alone does not elevate a claim. Meaning is not a physical force; coherence is not perfection or moral goodness; wisdom is not optimization; dual aspect is not proof of two substances or a completed theory of consciousness; the Spirit Field remains symbolic language for lived interior coherence rather than an established physical field; recursion or similarity across scale does not prove shared mechanism; and balance does not require equality. No coherence, awareness, meaning, vulnerability, or responsibility measure determines human worth or exhausts the lived person.*

CHAPTER TWENTY-ONE

# MEANING: VALUE, PURPOSE, SEMANTICS, AND THE MATHEMATICS OF ORIENTATION

*The Twelfth Mathematical Chamber*

**Knowledge explains what is.**
**Meaning begins asking what should matter—and how understanding should now guide participation.**

## Guiding Question

How does organized knowledge become meaningful orientation, how can semantics, value, preference, purpose, agency, narrative, commitment, shared significance, and responsible participation be represented mathematically, and what distinguishes genuine meaning from information, reward, utility, habit, social approval, emotional intensity, ideological certainty, or efficient pursuit of a destructive goal?

Chapter Ten established state.

Chapter Eleven established relation.

Chapter Twelve established differentiation.

Chapter Thirteen established boundary.

Chapter Fourteen established vulnerability.

Chapter Fifteen established adaptation.

Chapter Sixteen established attention.

Chapter Seventeen established awareness.

Chapter Eighteen established learning.

Chapter Nineteen established memory.

Chapter Twenty established knowledge.

A system can now:

- occupy a state;
- participate through relationships;
- preserve an identity;
- encounter vulnerability;
- adapt;
- select information;
- experience;
- learn;
- remember;
- organize evidence into knowledge.

Yet knowledge alone does not determine orientation.

A person may know how a weapon functions without knowing whether it should be built.

A government may know how to influence behavior without possessing the right to manipulate it.

A physician may know the statistical probability of an outcome while still needing to understand what dignity requires in the particular life before them.

A civilization may know how to extract resources without knowing what it owes the generations that will inherit the consequences.

A machine may optimize a reward while remaining indifferent to everything the reward omitted.

Book I established this threshold through a constitutional recognition:

**Meaning is organized understanding that faithfully orients participation within reality.**

It further distinguished the roles of knowledge and meaning:

**Knowledge explains. Meaning directs.**
**Knowledge reveals. Meaning responds.**

Book I also introduced the symbolic relation:

$$Me = f(K, R, P)[ARC]$$

where:

- $Me$ represents meaningful orientation;
- $K$ represents organized knowledge;
- $R$ represents lived relationships;
- $P$ represents responsible participation.

It explicitly presented this relation as an architectural recognition rather than an empirical psychological law. It also introduced **Scientific Spiritualism** as disciplined reflection upon interior human existence after science has honestly spoken, and the **Spirit Field** as symbolic language for the lived integration of understanding, gratitude, hope, responsibility, forgiveness, wonder, purpose, and compassion.

Book II preserves that inheritance.

It now gives mathematical passports to the structures that can be measured without pretending that measurement alone creates meaning.

The governing transition is:

$$\text{Knowledge} \to \text{Valuation} \to \text{Orientation} \to \text{Commitment} \to \text{Participation} \to \text{Consequence} \quad [SDAE]$$

Every arrow requires discipline.

Knowledge may be incomplete.

Valuation may be distorted.

Orientation may be inherited without examination.

Commitment may become rigidity.

Participation may harm others.

Consequences may reveal that the entire goal architecture was mistaken.

Meaning therefore cannot be established merely because a system:

- pursues a target;
- maximizes utility;
- repeats a narrative;
- obeys a command;
- receives reward;
- experiences emotional intensity;
- belongs to a group;

- speaks confidently about purpose.

The deeper question is:

**What has become significant, why has it become significant, how does that significance orient action, and does the resulting participation remain faithful to reality, dignity, relationship, and consequence?**

## Terminological Discipline

The word **meaning** carries several distinct traditions. Book II must not allow them to collapse into one another.

| Term | Meaning in Book II |
|---|---|
| **Semantic meaning** | The relation through which an expression refers, denotes, describes, or contributes to truth conditions under an interpretation and context |
| **Speaker meaning** | What a communicator intends another participant to recognize through an utterance or signal |
| **Pragmatic meaning** | Meaning arising through context, intention, presupposition, conversational inference, and use |
| **Significance** | The experienced or normatively assigned importance of an event, being, relationship, commitment, or life |
| **Meaning in life** | A person's experienced comprehensibility, direction, significance, or related forms of existential orientation |
| **Meaning of life** | A broader philosophical or theological question concerning the significance or purpose of life as a whole |
| **Coherence of life** | Degree to which events and experiences form an intelligible pattern; distinct from SDAE coherence as the broader preservation of organized relationships |
| **Purpose** | A future-oriented aim, direction, or organizing commitment |
| **Goal** | A represented target state, region, outcome, or relationship toward which behavior is directed |
| **Terminal goal** | A goal pursued as an end rather than merely as a means |
| **Instrumental goal** | A goal pursued because it contributes to another goal |
| **Value** | A criterion of importance, desirability, worth, obligation, or priority |
| **Valuation** | A process assigning comparative importance to possibilities |
| **Intrinsic value** | Value attributed to something in itself rather than solely through what it produces |
| **Instrumental value** | Value attributed through contribution to another end |
| **Preference** | A comparative ordering among alternatives |
| **Utility** | A mathematical representation of preference or consequence under a declared decision model |
| **Welfare** | A represented account of how well a person or population is doing under declared dimensions |

| Term | Meaning in Book II |
|---|---|
| **Reward** | A numerical signal used by a learning or control system to alter behavior |
| **Cost** | A numerical representation of resource use, loss, burden, risk, or undesired consequence |
| **Objective function** | A quantity a mathematical procedure is designed to maximize or minimize |
| **Proxy** | A measurable quantity used in place of a harder-to-measure purpose |
| **Norm** | A rule, expectation, or standard governing behavior |
| **Obligation** | A requirement that is not exhausted by personal preference |
| **Dignity** | The standing of a person that must not be reduced to usefulness, productivity, compliance, or numerical score |
| **Flourishing** | Multidimensional development and participation extending beyond momentary pleasure or preference satisfaction |
| **Agency** | Capacity to initiate, select, or govern action under relevant constraints |
| **Autonomy** | Self-endorsed participation rather than mere absence of external influence |
| **Intention** | A represented commitment to act or bring about a condition |
| **Commitment** | A retained orientation constraining future action beyond immediate impulse |
| **Function** | A contribution performed by a component within a system or a role explained through a declared biological, technical, or social account |
| **Teleology** | Explanation in terms of ends or purposes |
| **Teleonomy** | Goal-directed or functionally organized biological behavior explained without assuming conscious foresight |
| **Narrative** | An ordered and interpreted relationship among events |
| **Narrative identity** | An evolving story through which a person relates past, present, anticipated future, and moral agency |
| **Shared meaning** | Significance or interpretation coordinated among several participants |
| **Convention** | A mutually sustained regularity that helps solve a recurrent coordination problem |
| **Common knowledge** | A condition known by all relevant participants, known to be known by all, and so on through the formal hierarchy |
| **Collective purpose** | A goal architecture through which several participants coordinate contribution |
| **Mattering** | Experienced or recognized significance of a life or contribution |
| **Meaningful participation** | Action guided by understanding, value, relationship, responsibility, and consequence |
| **Coherent meaning** | Meaning that remains reality-accountable, relationally faithful, ethically constrained, revisable, and capable of guiding responsible participation |

These terms must remain distinct.

$$\text{Meaning} \neq \text{Information} \quad \text{Meaning} \neq \text{Emotional Intensity}$$
$$\text{Preference} \neq \text{Moral Worth} \quad \text{Utility} \neq \text{Dignity} \quad \text{Reward} \neq \text{Goodness}$$
$$\text{Goal-Directedness} \neq \text{Conscious Purpose}$$
$$\text{Narrative Coherence} \neq \text{Historical Accuracy}$$
$$\text{Shared Meaning} \neq \text{Universal Truth} \quad \text{Purpose} \neq \text{Guaranteed Benefit}$$

## Scientific and Mathematical Foundations

This chamber stands upon the inheritance of:

Semantics.

Pragmatics.

Linguistics.

Psychology.

Neuroscience.

Anthropology.

Narrative psychology.

Decision theory.

Game theory.

Control theory.

Economics.

Behavioral science.

Evolutionary biology.

Social-choice theory.

Ethics.

Information theory.

Shannon's mathematical theory of communication formalized entropy, coding, transmission, and uncertainty while deliberately excluding semantic significance from the engineering problem it addressed. This limitation was methodological rather than a claim that meaning does not exist. It establishes one of this chamber's central boundaries: a channel may transmit symbols reliably without determining what those symbols mean, whether they are true, or why they matter.

Grice's analysis of conversation distinguished what an expression conventionally says from what a speaker communicates through intention, context, and conversational inference. Communication therefore cannot always be reduced to literal decoding.

Von Neumann and Morgenstern developed an axiomatic representation of preference under risk, providing the foundation for expected-utility theory. Kahneman and Tversky later demonstrated systematic patterns of human choice that challenged expected utility as a complete descriptive model, particularly through reference dependence and asymmetric treatment of gains and losses. These theories describe decision architecture; neither defines moral worth or existential meaning.

Psychological research on meaning in life has increasingly distinguished three dimensions: comprehensibility or coherence, purpose or direction, and significance or mattering. The Meaning in Life Questionnaire separately measures the experienced presence of meaning and the search for meaning. These are self-reported psychological constructs, not direct measurements of objective worth.

Self-determination theory has proposed that autonomy, competence, and relatedness are central to understanding human motivation and the quality of goal pursuit. This is one empirically developed psychological framework, not a universal equation of every human life.

Narrative psychology has investigated life stories as evolving structures through which people organize autobiographical memory, identity, culture, and anticipated future. A coherent narrative may support orientation while still remaining vulnerable to omission, distortion, and inherited cultural assumptions.

Cybernetics and biology introduced disciplined ways of speaking about purposive organization without presuming supernatural or conscious foresight. Rosenblueth, Wiener, and Bigelow analyzed purpose through goal-directed behavior and feedback, while Mayr emphasized the need to distinguish immediate biological mechanisms from evolutionary explanations.

Social mathematics reveals another boundary. Nash formalized one axiomatic solution to bargaining. Arrow proved that, under a specified set of fairness and rationality requirements, no general rule can convert unrestricted individual rankings over three or more alternatives into a complete social ordering while satisfying all those requirements simultaneously. Shared purpose is therefore not obtained merely by adding individual preferences.

These inheritances reveal several different questions.

What does a symbol refer to?

What does a speaker intend?

What does an observer value?

Which outcome is preferred?

Which goal governs action?

Which values may be traded?

Which must remain inviolable?

What makes a life intelligible?

What makes a life matter?

How can several people coordinate meaning without erasing difference?

## Architectural Recognition

Only after the sciences, human sciences, philosophy, and mathematics have spoken does the Architecture introduce this chamber's central recognition:

**Meaning is organized significance through which knowledge, relationship, value, and responsibility orient participation toward a future judged worthy of becoming.**

A general meaning architecture may be represented as:

$$M_e = (K, R, V, G, N, A, C, E) [SDAE / MOD]$$

where:

- $K$ is organized knowledge;
- $R$ is the lived relational architecture;
- $V$ is the value architecture;
- $G$ is the goal and purpose architecture;
- $N$ is narrative and temporal continuity;
- $A$ is agency and available action;

- $C$ is context and constraint;
- $E$ is ethical and evidentiary accountability.

Meaning is not established merely because:

$$G \neq \varnothing .$$

A missile possesses a target.

A thermostat possesses a setpoint.

A parasite possesses evolved mechanisms that support reproduction.

A corporation may possess a growth objective.

A person may possess a goal imposed through fear.

Goal-directedness is only one component.

Coherent meaning requires the goal to remain accountable to:

- knowledge;
- relationship;
- dignity;
- consequence;
- consent;
- revisability;
- reality.

## Three Domains of Meaning

The word **meaning** should be typed before it is mathematized.

### Semantic Meaning

What does a word, symbol, sentence, gesture, or signal refer to or communicate?

### Functional Meaning

What role does a component, behavior, or process perform within a system?

## Existential and Normative Meaning

What gives a life, action, relationship, or commitment significance, direction, and worth?

These domains may interact.

They are not one mechanism.

A sentence can possess semantic meaning without being important.

A biological process can possess a function without being consciously intended.

A relationship can possess profound significance without being expressible in one proposition.

$$\text{Semantic Meaning} \neq \text{Biological Function} \neq \text{Existential Significance}$$

## Semantic Interpretation

Let:

$$L$$

be a language.

Let:

$$D$$

be a domain of possible referents.

Let:

$$I$$

be an interpretation.

Let:

$$c$$

be context.

For expression $e$, write:

$$[\![ e ]\!]^{I,c}$$

for its semantic value.

For a name $n$:

$$[\![ n ]\!]^{I,c} \in D.$$

For a proposition $\varphi$:

$$[\![ \varphi ]\!]^{I,c} \in \{0,1\}$$

under a classical truth-valued semantics.

A context-sensitive expression may satisfy:

$$[\![ e ]\!]^{I,c_1} \neq [\![ e ]\!]^{I,c_2}.$$

The word **here** changes referent when context changes.

The word **I** changes referent when speaker changes.

The sentence may remain syntactically identical.

Its interpreted content changes.

## Equation Passport 21.1

*Semantic Interpretation*

$$e \mapsto [\![ e ]\!]^{I,c}$$

**Status:** $[¿/MOD]$

**Expression:** $e$

**Interpretation:** $I$

**Context:** $c$

**Semantic value:** $[\![ e ]\!]^{I,c}$

**Possible semantic values:**

- object;

- property;
- relation;
- function;
- truth value;
- probability distribution;
- structured content.

**Required declarations:**

- language;
- domain;
- interpretation;
- context;
- type system;
- ambiguity rules.

**What it establishes:** A formal relationship between expression and interpreted content.

**What it does not establish:**

- truth of the expression;
- speaker intention;
- emotional significance;
- moral importance;
- experiential understanding.

**Architectural significance:** Symbols become semantically usable only through a disciplined relation among expression, interpretation, context, and reality.

## Compositional Meaning

A compositional architecture may satisfy:

$$[\![ f(e_1,\ldots,e_k) ]\!] = F_f([\![ ,e_1, ]\!] \ldots [\![ e_k ]\!]).$$

The meaning of the whole depends upon:

- meanings of the parts;
- the rule by which they are combined.

This makes generative language possible.

A finite vocabulary can participate in indefinitely many expressions.

Yet natural communication also depends upon:

- context;
- implication;
- irony;
- presupposition;
- shared history;
- gesture;
- tone.

Formal semantics organizes part of meaning.

Pragmatics organizes another.

## Information Without Meaning

Let $S$ be a source state and $M$ a transmitted message.

Mutual information is:

$$I(S;M)=\sum_{s,m} p(s,m)\log\frac{p(s,m)}{p(s)p(m)}.$$

A large value means that observing $M$ reduces uncertainty concerning $S$ under the declared probability model.

It does not establish that:

- the message is true;
- the message refers correctly;
- the receiver understands;
- the information matters;
- the communication is ethical.

A code can transmit falsehood reliably.

A surveillance system can extract information efficiently while violating dignity.

A random identifier can carry high mutual information about a person while carrying no human significance in itself.

Information Measures Distinguishability.
Meaning Requires Interpretation and Significance.

## Communication as Coordination

Let:

$$s \in S$$

be a world state.

Let:

$$m \in M$$

be a signal.

Let:

$$a \in A$$

be a receiver action.

A sender uses:

$$\sigma(m \mid s).$$

A receiver uses:

$$\rho(a \mid m).$$

The probability of action $a$ in state $s$ is:

$$P(a \mid s) = \sum_m \sigma(m \mid s)\rho(a \mid m).$$

Let:

$$u(s, a)$$

measure coordination success.

Expected success is:

$$J_{sig} = \sum_{s,m,a} p(s)\sigma(m \mid s)\rho(a \mid m)u(s,a).$$

A signaling convention emerges when sender and receiver mappings become sufficiently aligned to coordinate action. Lewis used recurrent coordination problems to analyze conventions of meaning.

## Equation Passport 21.2

*Signaling and Conventional Meaning*

$$J_{sig} = \sum_{s,m,a} p(s)\sigma(m \mid s)\rho(a \mid m)u(s,a)$$

**Status:** $[MOD]$

**World state:** $s$

**Signal:** $m$

**Action:** $a$

**Sender mapping:** $\sigma$

**Receiver mapping:** $\rho$

**Coordination value:** $u$

**Perfect deterministic convention:**

$$\rho(a^{¿}(s) \mid \sigma(s)) = 1$$

for every relevant state.

**Possible failures:**

- ambiguous signal;
- incompatible codebooks;
- deceptive sender;
- changed context;
- unequal incentives;
- inaccessible signal;
- coercive convention.

**What coordination does not establish:** Truth, justice, or subjective understanding.

**Architectural significance:** Shared semantic meaning can emerge when relationships among world, signal, interpretation, and action become mutually stabilized.

## Speaker Meaning

Let:

- $u$ be an utterance;
- $L(u)$ be its literal semantic content;
- $c$ be context;
- $i$ be speaker intention;
- $B_H$ be the hearer's background beliefs.

The interpreted message may be:

$$M_H = P(L,(u),c,i B_H)[MOD]$$

where $P$ is pragmatic inference.

A person saying:

It is cold in here.

may be:

- reporting temperature;
- requesting a window be closed;
- expressing discomfort;
- criticizing another person's action.

The words alone may not determine which.

Meaning therefore belongs partly to relationship.

The hearer must infer what the speaker is trying to make recognizable.

## Meaning in Life

Psychological research often separates three dimensions.

Let:

$$m_{life} = \begin{bmatrix} C_L \\ P_L \\ S_L \end{bmatrix} [¿/MOD]$$

where:

- $C_L$ is comprehensibility or life coherence;
- $P_L$ is purpose or direction;
- $S_L$ is significance or mattering.

## Comprehensibility

Does the life possess an intelligible pattern?

## Purpose

Does the life possess enduring direction?

## Significance

Does the life feel worthy, consequential, or valuable?

Martela and Steger argued that these dimensions should be distinguished rather than treated as interchangeable.

A life may possess:

- direction without intelligibility;
- intelligibility without mattering;
- mattering without one dominant goal;
- purpose inherited from others but not personally endorsed.

## Equation Passport 21.3

***Meaning-in-Life Signature***

$$m_{life} = (C_L, P_L, S_L)^T$$

**Status:** $[¿/MOD]$

**Comprehensibility:** $C_L$

**Purpose:** $P_L$

**Significance: $S_L$**

**Possible measurement sources:**

- self-report;
- interview;
- narrative analysis;
- longitudinal behavior;
- social and relational context.

**Required declarations:**

- population;
- cultural context;
- time horizon;
- instrument;
- construct validity;
- response conditions.

**What it does not measure directly:**

- objective moral worth;
- historical truth;
- social justice;
- clinical diagnosis;
- metaphysical purpose.

**Architectural significance:** Meaning in life is multidimensional; one strong dimension should not conceal the collapse of another.

## Presence and Search

Let:

$$M_P$$

represent the experienced presence of meaning.

Let:

$$M_S$$

represent active search for meaning.

These are not opposites.

A person may possess:

$$M_P \gg 0, M_S \gg 0.$$

Meaning may be present while still deepening.

A person may also report low present meaning and high search.

The Meaning in Life Questionnaire was designed to distinguish these dimensions rather than collapse them into one score.

Search can represent:

- growth;
- transition;
- curiosity;
- crisis;
- inherited meaning no longer fitting reality.

High search is not automatically health.

Low search is not automatically completion.

## Perceived Meaning and Coherent Meaning

A person may experience intense meaning within:

- a destructive ideology;
- an abusive relationship;
- a fraudulent movement;
- a violent cause;
- a false historical narrative.

Subjective meaning is real as subjective experience.

Its orientation may remain harmful.

The Architecture therefore distinguishes:

$$M_{perceived}$$

from:

$$M_{coherent}.$$

Coherent meaning requires additional accountability to:

- truth;
- dignity;
- consequence;
- consent;
- non-domination;
- revisability.

A Life Can Feel Meaningful While Being Misdirected.

## Value as a Relation

Value is often spoken of as though it were simply inside an object.

Yet valuation usually relates:

- an object or condition;
- a valuer;
- a context;
- a time;
- a consequence;
- a standard.

Let:

$$v_i(x;c;t;\Omega)$$

be the value assigned by participant $i$ to state $x$, in context $c$, at time $t$, under evaluative domain $\Omega$.

The same object may possess different instrumental value in different contexts.

Water may be ordinary near a river.

It may be decisive in a desert.

But relational valuation does not imply that every value is arbitrary.

Some conditions constrain valuation through reality.

A poison does not become nourishing because it is preferred.

A person does not lose dignity because an institution assigns a low score.

## Intrinsic and Instrumental Value

Let:

$$V_{\int ¿(x)¿}$$

represent value attributed to $x$ in itself.

Let:

$$V_{instr}(x \rightarrow y)$$

represent value attributed through contribution to $y$.

A medical instrument may possess primarily instrumental value.

A human life should not be reduced to:

$$V_{instr} = \text{economic productivity}.$$

A relationship may possess intrinsic significance while also producing instrumental benefits.

The distinction prevents means from silently replacing ends.

## Preference Relations

Let $X$ be a finite set of alternatives.

Write:

$$x \succeq y$$

when $x$ is weakly preferred to $y$.

Define strict preference:

$$x \succ y \iff x \succsim y \text{ and not } y \succsim x.$$

Define indifference:

$$x \sim y \iff x \succsim y \text{ and } y \succsim x.$$

If $\succsim$ is complete and transitive on a finite set, it can be represented by a utility function:

$$x \succsim y \iff u(x) \geq u(y).$$

The numerical values represent ordering.

They do not measure moral worth.

Any strictly increasing transformation:

$$\tilde{u} = \phi(u)$$

preserves the ordinal ranking.

## Equation Passport 21.4

*Ordinal Utility*

$$x \succsim y \iff u(x) \geq u(y)$$

**Status:** $[THM / MOD]$

**Alternatives:** $X$

**Preference relation:** $\succsim$

**Representation:** $u : X \to R$

**Core conditions in the finite prototype:**

- completeness;
- transitivity.

**Uniqueness:** Up to strictly increasing transformation

**What utility represents:** Comparative preference

**What utility does not represent automatically:**

- happiness;
- virtue;

- dignity;
- moral value;
- social welfare;
- meaning in life.

**Architectural significance:** Numerical representation can clarify choice without turning every object of choice into a commensurable measure of worth.

## Expected Utility

Let a lottery be:

$$L=\{(p_1,x_1),\ldots,(p_n,x_n)\}$$

with:

$$p_i\geq 0,\sum_i p_i=1.$$

Expected utility is:

$$U(L)=\sum_{i=1}^{n} p_i u(x_i).$$

Under the von Neumann–Morgenstern axioms for preferences over lotteries, such a representation exists, with $u$ unique up to positive affine transformation.

### Equation Passport 21.5

*Expected Utility*

$$U(L)=\sum_i p_i u(x_i)$$

**Status:** $[THM/MOD]$

**Outcomes:** $x_i$

**Probabilities:** $p_i$

**Outcome utility:** $u(x_i)$

**Lottery utility:** $U(L)$

**Required assumptions include:**

- complete and transitive preferences;

- continuity;
- independence;
- correctly represented probabilities.

**Positive affine invariance:**

$$\tilde{u} = au + b, a > 0.$$

**What it provides:** A disciplined normative representation of choice under risk.

**What it does not establish:** That observed human choice always follows the model or that utility equals moral significance.

**Architectural significance:** Decisions under uncertainty can be represented mathematically while preserving the boundary between preference consistency and ethical value.

## Reference-Dependent Valuation

Human choices are frequently evaluated relative to a reference point.

Let:

$$r$$

be the reference.

Let:

$$z_i = x_i - r.$$

A prospect-theoretic prototype is:

$$V(L; r) = \sum_i \pi(p_i) v(z_i) [MOD]$$

where:

- $v$ is a reference-dependent value function;
- $\pi$ is a decision-weighting function.

Kahneman and Tversky developed prospect theory after documenting patterns inconsistent with expected utility as a complete descriptive account, including sensitivity to gains and losses relative to a reference point.

### Equation Passport 21.6

*Reference-Dependent Choice*

$$V(L;r)=\sum_i \pi(p_i)\,v(x_i-r)$$

**Status:** $[MOD]$

**Reference:** $r$

**Gain or loss:** $x_i - r$

**Decision weight:** $\pi(p_i)$

**Subjective value:** $v$

**Possible descriptive features:**

- reference dependence;
- diminishing sensitivity;
- asymmetric gain–loss response;
- nonlinear probability weighting.

**What it does not establish:**

- what the reference ought to be;
- whether the preference is informed;
- whether the outcome is ethical;
- whether loss aversion should govern public policy.

**Architectural significance:** Meaning and value are influenced by inherited reference points; changing the reference can change experienced significance without changing the physical outcome.

## The Reference Point Can Be Harmful

A person accustomed to neglect may use neglect as the reference.

A community accustomed to exclusion may treat partial inclusion as extraordinary generosity.

A body accustomed to chronic strain may experience harmful conditions as normal.

A civilization accustomed to exploitation may define justice relative to an unjust baseline.

$$\text{Adaptation of the Reference} \not\Rightarrow \text{Health of the Reference}.$$

Meaning must therefore examine not only:

What do I value relative to my present expectations?

but:

Why did these expectations become the standard?

## Multiple Values

Human orientation ordinarily contains more than one objective.

Let:

$$J(a)=\begin{bmatrix} J_1(a) \\ J_2(a) \\ \vdots \\ J_m(a) \end{bmatrix}$$

represent several losses to be minimized.

They may include:

- harm;
- resource consumption;
- inequality;
- uncertainty;
- delay;
- ecological loss;
- loss of autonomy;
- loss of dignity.

Alternative $a$ Pareto-dominates $b$ when:

$$J_i(a) \leq J_i(b)$$

for every $i$, and:

$$J_j(a) < J_j(b)$$

for at least one $j$.

A Pareto-optimal alternative cannot be improved in one objective without worsening at least one other objective.

This does not identify one morally correct alternative.

It reveals the tradeoff surface.

## Equation Passport 21.7

*Multiobjective Value*

$$J(a) = (J_1,(a),\ldots J_m(a))^T$$

**Status:** $[¿/MOD]$

**Alternative:** $a$

**Objectives:** $J_i$

**Pareto dominance:**

$$a \prec_P b$$

when $a$ is no worse in every objective and better in at least one.

**Pareto frontier:** Set of nondominated alternatives

**What it reveals:** Tradeoffs among distinguishable values

**What it does not determine:**

- priority;
- fairness;
- rights;
- moral legitimacy;
- whose objectives belong in the vector.

**Architectural significance:** Meaning becomes more honest when values remain separately visible instead of being prematurely compressed into one score.

## Weighted Sums

A scalar objective may be:

$$J_w(a)=\sum_{i=1}^{m} w_i J_i(a)$$

where:

$$w_i \geq 0.$$

The weights encode priority.

They are not discovered by the optimization.

They are supplied.

Changing:

$$w$$

can change the preferred solution even when all empirical predictions remain unchanged.

The equation therefore contains value judgment before calculation begins.

Optimization Does Not Choose the Values.
Optimization Serves the Values It Was Given.

## Noncompensable Values

Some conditions should not be traded against aggregate benefit.

Let:

$$H_j(a) \leq 0$$

represent an inviolable constraint.

The optimization becomes:

$$\min_{a} J_w(a)$$

subject to:

$$H_j(a) \le 0 \; \forall j.$$

Possible noncompensable boundaries include:

- informed consent;
- prohibition of torture;
- protection from discrimination;
- minimum safety;
- due process;
- protection of children;
- respect for personhood.

A thousand small benefits should not automatically compensate for one violation of a constitutional boundary.

## Lexicographic Values

Suppose priorities are ordered:

$$J_1 \triangleright J_2 \triangleright \cdots \triangleright J_m.$$

Alternative $a$ is preferred lexicographically to $b$ when the first objective on which they differ favors $a$.

This models priorities such as:

1. avoid irreversible loss of life;
2. preserve dignity;
3. reduce suffering;
4. reduce cost.

The architecture rejects a monetary trade if a prior constitutional condition fails.

Lexicographic priority can also become rigid.

The order must remain ethically and empirically defensible.

## Values as a Motivational Structure

Schwartz and colleagues modeled basic human values as related positions in a motivational continuum, with some values compatible and others in tension. Their refined theory distinguished nineteen values across fifteen samples in ten countries while preserving the concept of an organized circular structure. It remains one psychological theory of reported value organization rather than a universal moral map.

This suggests an important architecture:

$$\text{Value Conflict} \neq \text{Absence of Values}.$$

A person may value:

- security;
- freedom;
- tradition;
- novelty;
- self-direction;
- care for others.

The difficulty is often not that nothing matters.

It is that several things matter simultaneously.

## Capabilities

Resources do not translate identically into real opportunity.

Let:

$$r_i$$

be resources available to person $i$.

Let:

$$z_i$$

represent personal, social, environmental, and institutional conversion conditions.

Let:

$$a \in A_i$$

be an available use of resources.

A capability set may be represented as:

$$C_i(r_i, z_i) = \{f, : f = \Phi(a; r_i; z_i) a \in A_i\} [MOD]$$

where $f$ represents achievable functionings.

Sen's capability approach challenged evaluations based only upon goods, utility, or subjective satisfaction by emphasizing what persons are actually able to be and do.

## Equation Passport 21.8

*Capability Set*

$$C_i(r_i, z_i) = \{f = \Phi(a; r_i; z_i) : a \in A_i\}$$

**Status:** $[MOD]$

**Resources:** $r_i$

**Conversion conditions:** $z_i$

**Available actions:** $A_i$

**Achievable functionings:** $f$

**Capability set:** $C_i$

**What it distinguishes:**

- possession of resources;
- practical freedom;
- achieved state;
- subjective satisfaction.

**Possible constraints:**

- disability;
- discrimination;
- geography;
- law;
- social role;
- environmental conditions;
- lack of knowledge.

**Architectural significance:** Meaningful agency requires more than formal permission; it requires real capacity to participate.

# Preference Under Constraint

Observed choice does not necessarily reveal unconstrained value.

Let:

$$a_i^{obs} = \arg \max_{a \in A_i^{available}} u_i(a).$$

The observed action depends upon:

- utility;
- available alternatives;
- knowledge;
- coercion;
- cost;
- fear;
- social pressure.

If:

$$A_i^{available}$$

is severely restricted, choosing $a$ does not mean $a$ was truly desired.

A person accepting dangerous work may value survival, not danger.

A patient consenting under misinformation may not have exercised informed choice.

$$\text{Chosen} \not\Rightarrow \text{Freely Endorsed}.$$

# Goal-Directed Control

Let:

$$x_t$$

be current state.

Let:

$$g_t$$

be a goal.

Let:

$$y_t = h(x_t)$$

be the observed output.

Define goal error:

$$e_t = g_t - y_t.$$

A controller selects:

$$u_t = \pi(e_t)$$

and the system evolves:

$$x_{t+1} = F(x_t, u_t, d_t).$$

This is mathematically purposive behavior in a control-theoretic sense.

The action changes in relation to departure from a goal.

## Equation Passport 21.9

***Goal-Directed Regulation***

$$e_t = g_t - h(x_t) u_t = \pi(e_t) x_{t+1} = F(x_t, u_t, d_t)$$

**Status:** $[MOD]$

**Goal:** $g_t$

**Observed state:** $h(x_t)$

**Error:** $e_t$

**Action:** $u_t$

**Disturbance:** $d_t$

**Goal-directed achievement:** Action varies so as to reduce or manage declared error.

**What it does not establish:**

- conscious intention;
- subjective purpose;
- moral value of the goal;
- wisdom of the controller;
- autonomy.

**Architectural significance:** Purpose can be represented behaviorally as organized reduction of departure, while lived purpose requires a deeper account of endorsement and significance.

## A Setpoint Is Not a Meaning

A thermostat can maintain temperature.

A missile can seek a target.

A biological pathway can stabilize a variable.

A person can organize decades of life around caring for a family, discovering truth, or serving a community.

These are all goal-related.

They are not experientially or ethically equivalent.

$$\text{Feedback Goal} \neq \text{Lived Purpose}.$$

Lived purpose ordinarily includes some combination of:

- awareness;
- temporal continuity;
- self-endorsement;
- relationship;
- significance;
- responsibility;
- interpretation.

## Goal Hierarchies

Let:

$$G=(V_G, E_G)$$

be a directed acyclic graph of goals.

An edge:

$$g_i \rightarrow g_j$$

means that achieving $g_i$ contributes instrumentally to $g_j$.

Terminal goals belong to:

$$T_G=\{g \in V_G : outdeg(g)=0\}.$$

A daily task may support a project.

The project may support a vocation.

The vocation may support service.

Service may support an understanding of a life worth living.

Purpose becomes intelligible when local action can be related to a larger architecture.

### Equation Passport 21.10

***Hierarchical Purpose***

$$G=(V_G, E_G) g_i \rightarrow g_j$$

**Status:** $[MOD/ARC]$

**Goal nodes:** $V_G$

**Support relations:** $E_G$

**Terminal goals:** $T_G$

**Instrumental goals:** $V_G \setminus T_G$

**Possible failures:**

- cyclic dependency;
- hidden terminal goal;
- means replacing ends;
- contradictory goals;
- inherited goal never examined;
- local success damaging global purpose.

**Architectural significance:** Meaningful action becomes possible when the immediate task remains related to the larger purpose it is meant to serve.

## Means Can Replace Ends

An institution may begin with the purpose:

$$\text{Educate Students}.$$

It develops metrics:

$$\text{Test Scores}.$$

The metric becomes the operating target.

The institution eventually teaches the test while impairing education.

The sequence is:

$$\text{Purpose} \to \text{Proxy} \to \text{Optimization} \to \text{Proxy Becomes Purpose}.$$

The same failure appears when:

- publication count replaces scientific understanding;
- revenue replaces service;
- compliance replaces trust;
- engagement replaces communication;
- survival replaces living.

## Temporal Purpose

Let:

$$u_t$$

represent value experienced or produced at time $t$.

A discounted-utility architecture is:

$$U_t = \sum_{\tau=t}^{T} \gamma^{\tau-t} u_\tau, 0 < \gamma \leq 1.$$

Samuelson's early utility work helped formalize intertemporal aggregation through discounting.

A smaller $\gamma$ reduces the present influence of distant consequences.

A larger $\gamma$ preserves them more strongly.

But no empirical measurement alone determines the ethically proper discount rate for:

- future generations;
- irreversible ecological loss;
- long-lived infrastructure;
- cultural inheritance.

## Equation Passport 21.11

*Intertemporal Orientation*

$$U_t = \sum_{\tau=t}^{T} \gamma^{\tau-t} u_\tau$$

**Status:** $[MOD]$

**Present time:** $t$

**Future value:** $u_\tau$

**Discount factor:** $\gamma$

**Time horizon:** $T$

**Low** $\gamma$**:** Strong present priority

**High** $\gamma$**:** Strong future inclusion

**Required declarations:**

- whose future;
- which consequences;
- uncertainty;

- reversibility;
- intergenerational standing.

**What discounting does not decide:** Whether a future person counts less morally merely because they live later.

**Architectural significance:** Purpose extends the present by permitting unrealized futures to influence present action.

## Commitment

A commitment constrains future choice.

Let current policy options be:

$$\Pi_t.$$

A commitment may restrict future policies to:

$$\Pi^C_{t+\tau} \subseteq \Pi_{t+\tau}.$$

This can protect:

- fidelity;
- scientific integrity;
- relational trust;
- long-term projects;
- constitutional rights.

Commitment can also preserve:

- obsolete doctrine;
- abusive loyalty;
- harmful identity;
- failed strategy.

Coherent commitment requires a revision condition.

Let:

$$R_C$$

be the evidence or ethical condition under which the commitment must be reconsidered.

A commitment with no possible revision becomes dogma.

A commitment revised by every impulse becomes no commitment at all.

## Self-Determined Purpose

Self-determination theory distinguishes the content of goals from the regulatory reasons through which those goals are pursued. It proposes autonomy, competence, and relatedness as important psychological needs affecting motivation and well-being.

A provisional purpose-quality vector is:

$$q_P = \begin{bmatrix} A_u \\ C_o \\ R_e \end{bmatrix} [MOD]$$

where:

- $A_u$ is autonomy or endorsement;
- $C_o$ is competence or effective capacity;
- $R_e$ is relatedness or meaningful connection.

A goal may be highly effective while not self-endorsed.

A goal may be self-endorsed while impossible under present capability.

A goal may be personally endorsed while relationally destructive.

Purpose therefore requires several dimensions.

## Reward and Purpose

In reinforcement learning, the return is:

$$G_t = \sum_{k=0}^{\infty} \gamma^k R_{t+k+1}.$$

A policy may seek:

$$\pi^{*} = \arg\max_{\pi} E_{\pi}[G_t].$$

The system is mathematically organized around reward.

This does not establish that reward equals:

- meaning;
- moral good;
- welfare;
- truth;
- human value.

A reward is an input to optimization.

Meaning asks whether the reward deserves authority.

## Inferring Values from Behavior

Suppose an observer sees policy:

$$\pi.$$

Inverse reinforcement learning seeks reward functions under which $\pi$ would be optimal or near-optimal.

Define:

$$R(\pi) = \{ r : \pi \in \arg\max_{\pi'} J_r(\pi') \}.$$

Ng and Russell showed that one observed policy may be compatible with many reward functions, making reward inference underdetermined without additional assumptions.

### Equation Passport 21.12

***Inverse Value Inference***

$$R(\pi) = \{ r : \pi \in \arg\max_{\pi'} J_r(\pi') \}$$

**Status:** $[MOD]$

**Observed policy:** $\pi$

**Candidate reward:** $r$

**Reward-compatible set:** $R(\pi)$

**Meaning:** Several distinct objectives may generate the same behavior.

**Sources of ambiguity:**

- constraint;
- habit;
- fear;
- incomplete knowledge;
- environmental limitation;
- imperfect optimization;
- hidden state.

**What behavior does not reveal uniquely:** The agent's true value structure.

**Architectural significance:** Meaning cannot be read directly from behavior without understanding the constraints, beliefs, history, and alternatives surrounding the behavior.

## Purpose in Machines

An artificial system may possess:

- an objective;
- a target;
- a reward;
- a planning horizon;
- a self-monitoring process;
- language describing purpose.

These support functional purposiveness.

They do not, by themselves, establish:

- lived significance;
- felt commitment;
- moral concern;
- awareness;
- human-like meaning.

A machine can optimize:

$$J$$

without caring about $J$.

The behavior may still carry immense consequence for beings who do care.

## Teleology and Teleonomy

Rosenblueth, Wiener, and Bigelow treated purposive behavior as behavior interpretable through relation to a goal, particularly when feedback changes action according to present deviation and anticipated consequence.

Biological purpose requires additional discipline.

Let:

$$M_z$$

represent the present mechanism of trait $z$.

Let:

$$O_z$$

represent its present organism-level effect.

Let:

$$H_z$$

represent its evolutionary or developmental history.

A teleonomic account may be represented as:

$$T_z = (M_z, O_z, H_z) [MOD]$$

The heart pumps blood through a present mechanism.

The explanation of why hearts possess this organization also involves biological history.

Mayr emphasized the distinction between immediate functional causes and evolutionary causes.

### Equation Passport 21.13

*Teleonomic Function*

$$T_z = (M_z, O_z, H_z)$$

**Status:** $[MOD]$

**Present mechanism:** $M_z$

**System contribution:** $O_z$

**Developmental or evolutionary history:** $H_z$

**What it permits:** Disciplined function language in biology

**What it does not require:**

- conscious foresight;
- cosmic intention;
- future causes acting backward;
- moral value.

**Failure condition:** A present effect is called the purpose of a trait without historical or mechanistic justification.

**Architectural significance:** Function can be represented without projecting human intention into every organized process.

## Not Everything Stable Has a Purpose

A star may remain in hydrostatic equilibrium.

A crystal may minimize free energy under specified conditions.

A river may form a channel.

These possess lawful organization.

They need not possess goals.

The Architecture may recognize:

- stability;
- relation;
- boundary;
- adaptation-like dynamics.

It must not silently import:

- intention;
- desire;
- purpose;
- meaning.

$$\text{Organization} \not\Rightarrow \text{Purpose}.$$

## Narrative Meaning

A life is not encountered only as a set of independent events.

Events become related through:

- temporal order;
- causal interpretation;
- value;
- identity;
- anticipated future.

Let:

$$E = \{e_1, \ldots, e_n\}$$

be remembered events.

Let:

$$\prec$$

be temporal order.

Let:

$$C_N$$

be causal or explanatory edges.

Let:

$$V_N$$

assign significance.

Let:

$$S_N$$

represent the continuing self.

A narrative is:

$$N = (E, <, C_N, V_N, S_N) [ARC / MOD]$$

McAdams described life stories as evolving narratives through which autobiographical memory, personality, identity, and culture become organized.

## Equation Passport 21.14

*Narrative Meaning*

$$N = (E, <, C_N, V_N, S_N)$$

**Status:** $[ARC / MOD]$

**Events:** $E$

**Temporal order:** $<$

**Causal interpretation:** $C_N$

**Value assignment:** $V_N$

**Continuing self:** $S_N$

**Possible narrative-quality dimensions:**

$$q_N = (Q_T, Q_C, Q_I, Q_E)$$

where:

- $Q_T$ is temporal organization;

- $Q_C$ is causal intelligibility;
- $Q_I$ is identity continuity;
- $Q_E$ is evidentiary fidelity.

**What narrative coherence does not establish:** Historical accuracy.

**Architectural significance:** Meaning can emerge when events become related to identity and future without allowing the story to replace the evidence.

## The Story Can Become a Prison

A person may inherit the narrative:

I always fail.

Every later event is interpreted through that story.

Success becomes an exception.

Ambiguity becomes evidence of failure.

The narrative acts as a prior.

Let:

$$p(e \mid N)$$

be the probability assigned to event interpretation $e$ under narrative $N$.

A rigid narrative may satisfy:

$$p(e_{confirm} \mid N) \gg p(e_{disconfirm} \mid N)$$

even when reality does not support the imbalance.

The story organizes experience.

It also filters correction.

## Narrative Truth

Narrative truth involves more than factual chronology.

It may organize:

- significance;
- responsibility;
- identity;
- transformation.

Yet a meaningful narrative should preserve distinctions among:

- what occurred;
- what was inferred;
- what was felt;
- what is now believed;
- what remains uncertain.

The statement:

This is what the event came to mean within my life

differs from:

This is exactly what every participant intended when it occurred.

## Redemptive Narratives

A person may construct meaning by interpreting suffering as contributing to later service, wisdom, or compassion.

This can be a genuine act of human reorganization.

It becomes harmful when imposed upon others.

No person should be told:

- the harm was necessary;
- the suffering was deserved;
- the injury became acceptable because growth followed;
- failure to find meaning is a moral weakness.

$$\text{Growth After Harm} \not\Rightarrow \text{Justification of Harm}.$$

# Meaning and Suffering

Frankl's logotherapy placed the search for meaning at the center of human orientation, including the possibility of responding meaningfully under conditions of suffering. His work should not be transformed into the claim that suffering is required, that preventable suffering should be accepted, or that those who did not survive lacked sufficient purpose.

The Architecture establishes four boundaries.

## Suffering Is Not Necessary for Meaning

Love.

Creation.

Learning.

Service.

Wonder.

Relationship.

These can become meaningful without injury.

## Preventable Suffering Should Not Be Romanticized

Meaning-making is not a substitute for:

- medical care;
- justice;
- safety;
- prevention;
- accountability.

## Meaning May Emerge in Response

A person may preserve dignity, responsibility, or love under conditions they did not choose.

### The Victim Does Not Owe the Observer a Lesson

Another person's pain should not be converted into inspiration without consent.

## Meaning and Mortality

Human beings can represent:

$$T_{life} < \infty .$$

Finitude introduces opportunity cost.

Choosing one path excludes another.

Committing time to one relationship leaves less time for another.

Mortality therefore changes the geometry of value.

Let:

$$B_T$$

be remaining time budget.

Let:

$$a_i$$

be time allocated to commitment $i$.

Then:

$$\sum_i a_i \leq B_T .$$

The equation does not determine how the life should be lived.

It reveals that meaning must eventually become allocation.

What receives time becomes part of the lived value architecture.

## Meaning and Attention

Chapter Sixteen established that attention allocates limited resources.

Meaning helps determine the priority scores that attention uses.

Let:

$$s_i = \alpha S_i + \beta R_i + \gamma U_i + \delta H_i - \lambda C_i.$$

Meaning influences:

- $R_i$: what is relevant;
- $H_i$: what harm matters;
- the weights themselves.

A person may say that family matters most while attention continually follows novelty.

This creates an alignment gap.

Let:

$$a_i$$

be actual attention and:

$$v_i$$

declared value.

A normalized mismatch may be:

$$D_{AV} = \frac{1}{2} \sum_i \left| \frac{a_i}{\sum_j a_j} - \frac{v_i}{\sum_j v_j} \right| . [ARC/MOD]$$

A large $D_{AV}$ indicates misalignment between declared importance and actual allocation.

It does not establish hypocrisy.

Constraints may explain the difference.

The mismatch invites examination.

## Meaning and Relationship

Book I established that meaning rarely exists in isolation and repeatedly emerges through participation:

- parents through children;
- teachers through students;
- scientists through discovery;
- physicians through healing;
- communities through service;
- civilizations through shared purpose.

The mathematical architecture is relational.

Let:

$$G_R = (V_R, E_R)$$

be a relational network.

Let:

$$w_{ij}$$

represent trust, responsibility, care, dependence, or another declared relationship.

Meaning associated with participant $i$ may depend upon:

$$M_i = F\left(x_i, \{w_{ij}\}_{j \in N(i)}, h_i, g_i\right).$$

A life's significance cannot always be inferred from isolated achievement.

It may reside in the relationships sustained.

## Shared Meaning

Let agents be:

$$1,\ldots,n.$$

Let:

$$\mu_i(s)$$

be agent $i$'s interpretation of symbol $s$.

Perfect interpretive agreement would require:

$$\mu_1(s)=\cdots=\mu_n(s).$$

Human shared meaning rarely requires perfect identity.

It may require sufficient compatibility for coordinated participation.

Define pairwise semantic distance:

$$d_{ij}(s)=d(\mu_i,(s)\mu_j(s)).$$

A shared-meaning condition may be:

$$d_{ij}(s)\leq\varepsilon$$

for all relevant participants and uses.

The tolerance $\varepsilon$ depends upon consequence.

Poetry permits broad interpretive variation.

A surgical instruction permits much less.

## Common Knowledge

Let:

$$K_i(E)$$

mean that agent $i$ knows event $E$.

Define mutual knowledge:

$$M_1(E)=\text{¿}\, i\, K_i(E).$$

Recursively:

$$M_{n+1}(E)=\text{¿}\, i\, K_i(M_n(E)).$$

Common knowledge is:

$$CK(E)=\text{¿}\, n=1\,\text{¿}\,\infty\, M_n(E).$$

Aumann formalized common knowledge and showed that, under shared-prior Bayesian assumptions, common knowledge of differing posterior probabilities cannot persist. The theorem depends upon strong assumptions and should not be converted into the claim that real people must agree whenever they are rational.

## Equation Passport 21.15

*Common Knowledge*

$$CK(E)=\text{¿}\, n=1\,\text{¿}\,\infty\, M_n(E)$$

**Status:** $[\text{¿}/THM]$

**Event:** $E$

**Individual knowledge:** $K_i(E)$

**Mutual knowledge:** $M_n(E)$

**Common knowledge:** $CK(E)$

**Possible roles in shared meaning:**

- convention;
- promise;
- law;
- ceremony;
- coordinated expectation;
- public commitment.

**What common knowledge does not establish:**

- truth of $E$;
- moral legitimacy;
- equal interpretation;
- absence of coercion.

**Architectural significance:** Shared meaning often requires not merely that participants know a commitment, but that they know others recognize the same public relationship.

## Bargaining and Shared Purpose

Let:

$$S \subseteq R^2$$

be feasible utility outcomes for two participants.

Let:

$$d=(d_1, d_2)$$

be the disagreement point.

The Nash bargaining solution is:

$$u^{*}=\arg\max_{u \in S,\, u_i \geq d_i} (u_1-d_1)(u_2-d_2).$$

Nash derived this solution from a particular set of axioms concerning bargaining. It is one mathematically distinguished solution, not a universal definition of justice.

### Equation Passport 21.16

*Bargained Shared Orientation*

$$u^{*}=\arg\max_{u \in S, u \geq d} \prod_i (u_i-d_i)$$

**Status:** $[THM/MOD]$

**Feasible outcomes:** $S$

**Disagreement point:** $d$

**Participant gains:** $u_i-d_i$

**Required declarations:**

- utility representation;
- feasible set;
- disagreement condition;

- bargaining power;
- information;
- consent.

**What the solution does not establish:**

- equal dignity;
- fair initial conditions;
- absence of coercion;
- comparability of utilities;
- moral legitimacy of $d$.

**Architectural significance:** Shared purpose is influenced by the alternatives available if agreement fails; unjust background conditions can produce formally voluntary but substantively coercive agreements.

## The Disagreement Point Contains Power

Suppose one participant's disagreement outcome is:

$$d_1 \ll d_2.$$

Participant 1 faces greater loss if negotiation fails.

The resulting agreement may maximize the bargaining product and still reflect severe structural inequality.

Meaningful agreement requires more than procedural consent.

It requires examination of:

- who can walk away;
- who controls resources;
- whose needs are urgent;
- whose alternatives were constrained before bargaining began.

## Aggregating Values

Let each person $i$ possess preference order:

$$\succeq_i.$$

A social welfare rule seeks:

$$F:\left(\gtrsim_1,\ldots,\gtrsim_n\right)\mapsto\gtrsim_S.$$

Arrow's theorem shows that, for three or more alternatives and unrestricted individual rankings, no aggregation rule simultaneously satisfies the specified conditions of collective rationality, unanimity, independence of irrelevant alternatives, and non-dictatorship.

## Equation Passport 21.17

*Social-Meaning Aggregation*

$$F:\left(\gtrsim_1,\ldots,\gtrsim_n\right)\to\gtrsim_S$$

**Status:** $[THM/MOD]$

**Individual rankings:** $\gtrsim_i$

**Social ranking:** $\gtrsim_S$

**Arrow conditions in the classical setting include:**

- unrestricted domain;
- social-ordering consistency;
- unanimity or Pareto responsiveness;
- independence of irrelevant alternatives;
- non-dictatorship.

**Theorem:** No general aggregation rule satisfies all simultaneously for at least three alternatives.

**What it does not establish:** That collective choice is impossible.

**What it establishes:** Some desirable condition must be restricted, weakened, or supplemented.

**Architectural significance:** Shared meaning cannot be manufactured by pretending that plural values combine without constitutional tradeoffs.

# Consensus Is Not Shared Meaning

Consensus may arise through:

- genuine agreement;
- exhaustion;
- fear;
- exclusion;

- conformity;
- information asymmetry;
- domination.

A group reporting:

$$100\,\%$$

agreement does not reveal how the agreement was produced.

Shared meaning should preserve:

- participation;
- dissent;
- comprehension;
- consent;
- revision.

## Meaning and Culture

Culture organizes:

- symbols;
- stories;
- ceremonies;
- values;
- roles;
- temporal memory;
- collective identity.

Let:

$$C_t = \left( \Sigma_t, N_t, V_t, R_t, I_t \right)$$

where:

- $\Sigma_t$ is the symbol system;
- $N_t$ is the narrative inheritance;
- $V_t$ is the value architecture;
- $R_t$ is the relational structure;
- $I_t$ is the institutional structure.

Cultural meaning is transmitted through:

$$C_{t+1} = T_t(C_t, E_t, Q_t)$$

where:

- $E_t$ is new experience;
- $Q_t$ is correction and questioning.

Culture is not merely repeated.

It is inherited, interpreted, and revised.

## Meaning and Ancestral Continuity

Ancestral meaning may be preserved through:

- language;
- name;
- measure;
- craft;
- landscape;
- spiritual practice;
- architecture;
- story;
- family;
- historical record.

The Architecture retains the ethics already established:

**Restoration without resentment.**

**Reverence without fabrication.**

**Continuity without anachronism.**

Meaning can be inherited without claiming that present interpretations are identical to ancient ones.

$$\text{Resonance} \neq \text{Historical Identity}.$$

The strongest ancestral meaning is not the one requiring the most extravagant claim.

It is the one capable of surviving historical discipline.

## Meaning and Social Power

Institutions influence meaning by determining:

- which identities are publicly named;
- which goals are rewarded;
- which histories are taught;
- which lives are represented as exemplary;
- which suffering is normalized;
- which forms of contribution are recognized.

Let:

$$V_i^{public}$$

be public valuation assigned to person or contribution $i$.

Let:

$$V_i^{actual}$$

represent a declared broader evaluation.

A public-value distortion is:

$$\Delta V_i = V_i^{public} - V_i^{actual}.$$

A civilization may systematically undervalue:

- caregivers;
- elders;
- children;
- disabled persons;
- unseen labor;
- ancestral communities;
- future generations.

Economic visibility is not equivalent to human significance.

## Meaning and Work

Work may become meaningful through:

- contribution;
- mastery;
- relationship;
- service;
- creativity;
- autonomy;
- participation in a valued purpose.

Work may become alienating when:

- the worker cannot identify the purpose;
- the output violates values;
- the person is treated as replaceable input;
- productivity erases dignity;
- livelihood depends upon silent harm.

A job can be well paid and meaningless.

A poorly paid role can be meaningful and unjustly compensated.

Meaning should never be used to excuse exploitation.

## Meaning and Care

Care often produces outcomes difficult to measure.

A caregiver may:

- reduce fear;
- preserve dignity;
- remain present;
- help another person feel recognized.

These may not appear fully in:

- revenue;
- throughput;
- productivity;
- procedure count.

A model that omits them does not prove they are valueless.

It proves the model is incomplete for that purpose.

## Meaning and Science

Science can reveal:

- consequences;
- mechanisms;
- probabilities;
- constraints;
- causal relationships.

Science alone does not select every value.

Yet values cannot responsibly ignore science.

The relationship is:

$$\text{Evidence Constrains Responsible Meaning.}$$
$$\text{Meaning Orients the Use of Evidence.}$$

Meaning without evidence becomes vulnerable to fantasy.

Evidence without meaning becomes vulnerable to misuse.

## The Is–Ought Boundary

Let:

$$P$$

be a set of descriptive premises.

Let:

$$N$$

be a normative conclusion.

A purely deductive movement:

$$P \vdash N$$

requires at least one normative bridge premise if $P$ contains only descriptive claims.

For example:

$$\text{Action } a \text{ causes preventable harm}$$

does not alone yield:

$$\text{Do not perform } a$$

without some premise concerning the moral relevance of preventable harm.

The normative bridge should be explicit.

Hidden values do not disappear because they were omitted from the equation.

## Meaning and Dignity

Let:

$$U_i(a)$$

be utility associated with treatment of person $i$.

A utilitarian aggregation might consider:

$$\sum_i U_i(a).$$

Dignity introduces constraints such as:

$$H_i(a) \leq 0$$

for every person $i$.

The constraint prevents one person's fundamental standing from being erased by aggregate gain.

Dignity is therefore not merely another small weight.

It can function constitutionally.

## Meaning and Agency

A meaningful choice requires more than two visible buttons.

Let:

$$A_i$$

be the formally offered actions.

Let:

$$A_i^{real} \subseteq A_i$$

be actions genuinely available after accounting for:

- knowledge;
- coercion;
- cost;
- physical capacity;
- social penalty;
- fear;
- time.

Agency is constrained by:

$$|A_i^{real}|.$$

But number of options alone is insufficient.

A hundred trivial options do not compensate for absence of one indispensable option.

## Meaning and Autonomy

Autonomy does not require that a person be uninfluenced.

No person develops outside relationship.

Autonomy concerns whether influence becomes:

- understood;
- integrated;
- reflectively endorsed;
- revisable.

A person may freely embrace an inherited commitment.

A person may also obey a preference implanted through coercive control.

The external behavior may appear identical.

The internal regulatory history differs.

## Meaning and Future Generations

Let generation $g$ choose action:

$$a_g.$$

Its consequences extend:

$$x_{g+1} = F(x_g, a_g).$$

Future generations do not participate in the present bargaining table.

Yet they inherit:

- ecological conditions;
- debt;
- institutions;
- technology;
- knowledge;
- risk;
- culture.

A coherent value architecture should therefore include:

$$V_{future}$$

despite the absence of present political power.

Meaning enlarges the constituency of responsibility beyond those able to reward us now.

## Meaning and the Cathedral Equation

The Cathedral Equation will be developed fully in the next chamber.

Meaning enters it architecturally by changing what the system attempts to generate and prevent.

Let:

$$C(t)$$

represent a declared coherence state.

A meaning-sensitive architecture may be:

$$\frac{dC}{dt}=G(C,\Pi,M,e,et)-L(C,d,\Pi,M,e,et)[ARC/MOD]$$

where:

- $Me$ is meaningful orientation;
- $\Pi$ is participation policy;
- $G$ represents coherence-generating processes;
- $L$ represents coherence-loss processes.

Meaning is not inserted as a new physical force.

It influences:

- goals;
- attention;
- policy;
- sacrifice;
- interpretation;
- commitment.

## Operational Descendant of $Me=f(K,R,P)$

Book I introduced:

$$Me=f(K,R,P).$$

To avoid confusing $P$ for participation with $P_L$ for purpose, Book II writes responsible participation as:

$$\Pi .$$

An expanded architecture is:

$$Me_t = F(K_t, R_t, \Pi_t, V_t, G_t, N_t, C_t)[ARC/MOD]$$

where:

- $K_t$ is organized knowledge;
- $R_t$ is lived relationship;
- $\Pi_t$ is responsible participation;
- $V_t$ is value structure;
- $G_t$ is the goal hierarchy;
- $N_t$ is narrative continuity;
- $C_t$ is context and constraint.

A meaning-quality signature is:

$$q_{Me} = \begin{bmatrix} C_L \\ P_L \\ S_L \\ A_R \\ F_R \\ E_R \end{bmatrix} [ARC/MOD]$$

where:

- $C_L$ is comprehensibility;
- $P_L$ is purpose;
- $S_L$ is significance;
- $A_R$ is alignment with reality;
- $F_R$ is relational fidelity;
- $E_R$ is ethical responsibility.

### Equation Passport 21.18

*Operational Meaning Architecture*

$$Me_t = F(K_t, R_t, \Pi_t, V_t, G_t, N_t, C_t) q_{Me} = (C_L, P_L, S_L, A_R, F_R, E_R)^T$$

**Status:** $[ARC/MOD]$

**Inherited variables from Book I:**

- knowledge;
- relationship;
- participation.

**Additional Book II variables:**

- values;
- goals;
- narrative;
- context;
- reality alignment;
- ethical accountability.

**What the vector prevents:** Premature collapse of meaning into one score

**Possible noncompensable failures:**

- falsehood;
- coercion;
- dehumanization;
- preventable harm;
- denial of agency;
- complete immunity from correction.

**Non-claim:** This is not a universal empirical law of psychology or spirituality.

**Architectural significance:** Meaning becomes coherent when knowledge, relationship, value, narrative, and action remain aligned without erasing uncertainty or dignity.

## Meaning Cannot Be Maximized Without Qualification

Suppose one defines:

$$M e^{\text{¿}} = \max_{\pi} Me(\pi).$$

What prevents the optimizer from producing:

- ideological certainty;
- emotional intensity;
- total group conformity;
- permanent commitment;
- suppression of doubt?

If those correlate with the score, the optimizer may increase them.

A meaning metric must therefore remain bounded by:

- truth;
- freedom;
- plurality;
- dignity;
- correction;
- uncertainty.

The goal is not maximum experienced meaning at any cost.

It is faithful orientation.

## Dual-Aspect Meaning

Let:

$$x_o(t)$$

represent exterior empirical state.

Let:

$$x_i(t)$$

represent modeled interior significance.

Let:

$$b(t)$$

represent the interface through which interior meaning becomes language, action, bodily response, and relationship.

Let:

$$v(t)$$

represent value structure.

Let:

$$g(t)$$

represent goals.

Let:

$$n(t)$$

represent narrative continuity.

A joint architecture is:

$$\dot{x}_o = f_o(x_o, a, e, d) \dot{x}_i = f_i(x_i, K, R, v, g, n)$$
$$\dot{b} = f_b(b, x_o, x_i, c)[ARC/MOD/CONJ]$$

Observations may include:

$$y_o = h_o(x_o, b) + \varepsilon_o$$

and:

$$y_i = h_i(x_i, x_o, b) + \varepsilon_i,$$

where $y_i$ may include report, narrative, commitment, and choice.

## Equation Passport 21.19

*Dual-Aspect Meaning Architecture*

$$\xi=(x_o,x_i,b,v,g,n)\dot{\xi}=F(\xi,K,R,a,e,c)$$

**Status:** $[ARC/MOD/CONJ]$

**Exterior consequence:** $x_o$

**Interior significance:** $x_i$

**Interface:** $b$

**Values:** $v$

**Goals:** $g$

**Narrative:** $n$

**Required empirical work:**

- operational definitions;
- report validation;
- longitudinal measurement;
- behavioral consequence;
- cultural interpretation;
- competing models.

**Non-claims:**

- no separate immaterial substance;
- no new physical force;
- no reduction of lived meaning to one neural state;
- no claim that all meaningfulness is verbally reportable.

**Architectural significance:** Meaning is lived internally and expressed externally through one participating life without making interior and exterior descriptions interchangeable.

# Scientific Spiritualism

Book I defines **Scientific Spiritualism** as disciplined reflection upon the interior dimensions of human existence in light of expanding scientific understanding.

It does not replace science.

It begins after science has spoken.

Its constitutional order is:

$$\text{Reality First} \rightarrow \text{Science} \rightarrow \text{Reflection} \rightarrow \text{Responsible Participation}\,[ARC]$$

Questions of:

- gratitude;
- responsibility;
- forgiveness;
- hope;
- beauty;
- wonder;
- purpose;
- compassion

are not answered exhaustively by one laboratory measurement.

Neither should they be protected from everything science reveals.

Scientific Spiritualism therefore rejects both:

$$\text{Spiritual Claim Without Evidentiary Humility}$$

and:

$$\text{Measurement Used to Erase Interior Life}.$$

## The Spirit Field and Meaning

The Spirit Field remains:

$$S_f\,[SYM/ARC]$$

as symbolic language for organized interior coherence.

A symbolic meaning state may be represented as:

$$S_f^{Me} = O(U, G, H, R, F, W, P, C)\,[SYM/ARC]$$

where:

- $U$ is understanding;

- $G$ is gratitude;
- $H$ is hope;
- $R$ is responsibility;
- $F$ is forgiveness;
- $W$ is wonder;
- $P$ is purpose;
- $C$ is compassion.

This may support:

- reflection;
- teaching;
- ceremony;
- journaling;
- ethical examination;
- spiritual practice.

It does not establish:

- a physical field;
- a measurable energy;
- a quantum force;
- inherited supernatural information;
- a clinical treatment;
- empirical proof of metaphysical claims.

The symbolic status is not a weakness.

It is the boundary preserving intellectual integrity.

## Gratitude

Gratitude can orient attention toward:

- inheritance;
- contribution;
- dependence;
- gift;
- relationship.

It becomes distorted when used to silence legitimate complaint.

A person may be grateful and still seek justice.

## Hope

Hope represents future possibility influencing present action.

Let:

$$F_t$$

be the represented future set.

Let:

$$p_t(f)$$

be believed possibility.

Let:

$$v_t(f)$$

be future value.

A symbolic hope measure may involve:

$$H_t = \sum_{f \in F_t} p_t(f) v_t(f) [ARC/MOD]$$

but hope is not reducible to optimistic probability.

Hope may remain when probability is small because commitment preserves the worth of trying.

False hope ignores evidence.

Despair treats uncertainty as certainty of failure.

## Forgiveness

Forgiveness should not be reduced to:

- forgetting;
- reconciliation without safety;
- removal of accountability;
- restoration of access;
- denial of injury.

Let:

$$m_h$$

be memory of harm.

Let:

$$\alpha_t$$

be the degree to which the memory governs present action.

Forgiveness may alter:

$$\alpha_t$$

without requiring:

$$m_h = 0.$$

Meaning changes the relation between past and future.

The historical truth remains.

## Wonder

Wonder begins when knowledge enlarges rather than closes reality.

A phenomenon can be explained and remain astonishing.

Scientific understanding does not reduce wonder unless wonder depended upon ignorance.

The Architecture recognizes:

$$\text{Explanation} + \text{Wonder}$$

as compatible.

## Purpose

Purpose becomes coherent when it:

- reflects reality;
- preserves dignity;
- remains relational;
- survives difficulty without becoming fanatical;
- remains open to correction;
- contributes beyond immediate self-interest.

## Compassion

Compassion gives another being's vulnerability significance within one's own action architecture.

Let:

$$L_j(a)$$

be loss experienced by another participant $j$.

A purely self-centered objective omits it.

A compassionate objective includes:

$$J_i(a) = J_i^{self}(a) + \lambda_{ij} L_j(a),$$

where:

$$\lambda_{ij} > 0.$$

But compassion is not one coefficient.

The coefficient illustrates that another's condition has entered the decision.

## The Bridge Species as Meaning-Builder

The Bridge Species can represent:

- what is;
- what was;
- what might be;
- what should not become;
- what another person may experience;
- what future generations may inherit;
- what values were hidden inside an objective;
- what purpose once governed an institution;
- whether that purpose still deserves authority.

This creates a distinctive recursive capacity.

Humanity can revise:

- actions;
- models;
- goals;
- value hierarchies;
- narratives;
- institutions;
- meanings.

A system that revises only means becomes more efficient.

A system that can revise ends becomes reflectively responsible.

## Meaning and Human Becoming

Human becoming is not merely expansion of capacity.

Capacity can enlarge harm.

Human becoming requires increasing alignment among:

$$\text{Knowledge}\ \text{Value}\ \text{Character}\ \text{Participation}\ \text{Consequence}.$$

Meaning becomes the compass through which understanding begins guiding life.

## The Ethical Boundary of Meaning

A person may not impose meaning upon another person's suffering.

A government may not define citizens' meaning as obedience.

A corporation may not call overwork meaningful in order to avoid fair compensation.

A religious authority may not call abuse sacred.

A scientific authority may not call measurement complete merely because interior significance is difficult to quantify.

Meaning must remain:

- offered rather than coerced;
- interpreted rather than dictated;
- relational rather than extractive;
- reality-accountable rather than immune from evidence.

## Common Failures of Meaning Architecture

### Meaning–Information Collapse

A signal's uncertainty reduction is treated as semantic or existential meaning.

### Semantic–Existential Collapse

The meaning of a sentence is confused with the meaning of a life.

### Syntax–Semantics Collapse

Well-formed symbols are treated as interpreted understanding.

### Literal–Pragmatic Collapse

The words spoken are treated as the entirety of what was communicated.

### Convention–Truth Collapse

A shared usage is treated as correspondence with reality.

### Coordination–Justice Collapse

Successful coordination is treated as ethical legitimacy.

## Goal–Purpose Collapse

A represented target is treated as lived purpose.

## Purpose–Goodness Collapse

Strong direction is treated as moral value.

## Function–Intention Collapse

A biological contribution is described as consciously intended.

## Teleonomy–Cosmic Teleology Collapse

Evolved organization is converted into proof of universal purpose.

## Value–Preference Collapse

What a person prefers is treated as what is morally valuable.

## Preference–Choice Collapse

Constrained behavior is treated as freely revealed preference.

## Utility–Worth Collapse

A numerical ranking is treated as human dignity.

## Reward–Goodness Collapse

An optimization signal is treated as moral truth.

## Proxy–Purpose Collapse

The measurable target replaces the underlying reason for measuring it.

## Weighted-Sum Concealment

Ethical priorities enter as weights without disclosure.

### Compensability Error

A severe rights violation is treated as compensable by enough aggregate benefit.

### Pareto–Justice Collapse

Nondominance is treated as fairness.

### Bargaining–Consent Collapse

Agreement under unequal exit options is treated as free consent.

### Consensus–Meaning Collapse

Public agreement is treated as genuine shared significance.

### Social-Aggregation Innocence

Collective preference is assumed to emerge without constitutional tradeoffs.

### Capability–Resource Collapse

Formal possession is treated as practical freedom.

### Autonomy–Isolation Collapse

Self-determination is treated as freedom from all relationship.

### Meaning–Happiness Collapse

Pleasure or positive affect is treated as complete meaning.

### Meaning–Well-Being Collapse

A self-report score is treated as the whole of flourishing.

### Presence–Search Confusion

Searching for meaning is treated as proof that meaning is absent.

## Comprehensibility–Truth Collapse

A story that makes sense is treated as historically accurate.

## Narrative–Event Collapse

The interpretation of an event is treated as the event itself.

## Redemptive Coercion

A person is required to find growth in suffering.

## Suffering Romanticism

Preventable harm is treated as spiritually necessary.

## Mortality Manipulation

Finitude is used to pressure immediate commitment.

## Commitment–Dogma Collapse

Endurance of purpose is treated as immunity from correction.

## Revision–Faithlessness Collapse

Changing a purpose after stronger evidence is treated as betrayal.

## Cultural Meaning–Universality Collapse

One community's symbolic order is imposed upon every life.

## Tradition–Truth Collapse

Age is treated as proof.

## Novelty–Truth Collapse

Newness is treated as superiority.

### Ancestral Fabrication

Desired continuity replaces historical evidence.

### Resonance–Identity Collapse

Similarity across eras is treated as exact transmission.

### Public Visibility–Significance Collapse

What receives attention is treated as what matters.

### Economic Value–Human Value Collapse

Market price is treated as dignity.

### Productivity–Purpose Collapse

Output becomes the measure of a life.

### Care Erasure

Relational presence is omitted because it is difficult to quantify.

### Scientific Reductionism

Measurable correlates are treated as the whole of meaning.

### Spiritual Immunity

Interior significance is protected from every empirical correction.

### Spirit-Field Physicalization

Symbolic interior coherence is described as a physical force without evidence.

### Machine-Purpose Anthropomorphism

An objective function is treated as felt purpose.

### Machine-Meaning Denial of Consequence

A system's lack of demonstrated experience is used to ignore the effects it has upon experiencing beings.

### Value Inference Overconfidence

Behavior is treated as uniquely revealing what a person values.

### Future-Generation Erasure

Only presently represented preferences enter the objective.

### Dignity as a Small Weight

Personhood is treated as one minor term among many compensable outcomes.

## The Meaning Constitution

Before any meaning claim enters Book II, it must answer:

1. Which domain of meaning is being discussed?
2. Is the claim semantic, pragmatic, functional, existential, narrative, social, spiritual, or ethical?
3. What expression, event, relationship, or life is being interpreted?
4. What interpretation governs the symbol?
5. What context governs the interpretation?
6. What domain contains the referents?
7. What truth conditions apply?
8. Is the expression ambiguous?
9. Is the ambiguity resolved?
10. Is literal meaning being distinguished from speaker meaning?
11. What intention is attributed to the speaker?
12. What evidence supports that attribution?
13. What background knowledge is assumed?
14. What conversational inference is required?
15. Is a signal being transmitted?
16. What world state does it distinguish?
17. What sender mapping is used?
18. What receiver mapping is used?
19. What coordination payoff is used?
20. Does successful coordination imply truth?
21. Can deception produce the same coordination?
22. What information quantity is measured?
23. Is mutual information being confused with semantic significance?

24. Does the message refer accurately?
25. Is the message true?
26. Is the receiver capable of understanding it?
27. Does the message matter to the receiver?
28. Is meaning in life being measured?
29. Is the question about comprehensibility?
30. Is it about purpose?
31. Is it about significance?
32. Are those dimensions being collapsed?
33. Is presence of meaning distinguished from search for meaning?
34. Is a self-report instrument used?
35. What exactly does the instrument measure?
36. What culture and language shaped the instrument?
37. Is subjective meaning being treated as objective moral worth?
38. Could the experienced meaning be harmful?
39. What makes the orientation reality-accountable?
40. What makes it ethically accountable?
41. What object or state is valued?
42. Who is doing the valuing?
43. In what context?
44. At what time?
45. Under which evaluative standard?
46. Is the value intrinsic or instrumental?
47. Could an instrumental means replace the terminal end?
48. What preference relation is used?
49. Is it complete?
50. Is it transitive?
51. Is utility ordinal or cardinal within the model?
52. What transformations preserve its meaning?
53. Is expected utility used?
54. What axioms are assumed?
55. Are the probabilities known?
56. Are they subjective?
57. Is the model normative or descriptive?
58. Is reference dependence present?
59. What reference point is used?
60. Who established the reference?
61. Could the reference itself be harmful?
62. Are gains and losses treated differently?
63. What decision-weighting model is used?
64. Are several values present?
65. What is the objective vector?
66. Which alternatives are Pareto-dominated?
67. Which are Pareto-efficient?
68. Is Pareto efficiency being confused with justice?
69. Is a weighted sum used?
70. Who selected the weights?
71. Are the weights public?
72. Are the units commensurable?
73. What values are noncompensable?
74. What constitutional constraints apply?

75. Is lexicographic priority used?
76. Why is the priority order justified?
77. What capabilities are available?
78. What resources are available?
79. What conversion factors limit use?
80. Is formal choice being confused with real freedom?
81. What goal is represented?
82. Is the goal a point, region, trajectory, or relationship?
83. Is the goal terminal or instrumental?
84. What higher goal does it support?
85. Does the goal hierarchy contain cycles?
86. Has the means replaced the end?
87. What feedback compares state with goal?
88. Is the behavior merely regulatory?
89. Is conscious purpose being claimed?
90. What evidence supports conscious purpose?
91. Is the purpose self-endorsed?
92. Was the purpose inherited?
93. Was it imposed?
94. Was it adopted under coercion?
95. Does the participant possess real alternatives?
96. Does the purpose support autonomy?
97. Does it support competence?
98. Does it preserve relatedness?
99. Is a reward function used?
100. Who designed the reward?
101. Can the reward be exploited?
102. What value was omitted?
103. Is behavior being used to infer reward?
104. Could several reward functions explain the same behavior?
105. What constraints shaped the behavior?
106. Is a biological function being described?
107. What present mechanism produces it?
108. What contribution does it make?
109. What developmental or evolutionary history explains it?
110. Is conscious intention being projected onto biology?
111. Is purpose being projected onto nonliving stability?
112. What temporal horizon governs the purpose?
113. What discount factor is used?
114. Whose future is discounted?
115. Are future generations represented?
116. Is an irreversible loss discounted as though it were replaceable?
117. What commitment preserves the purpose?
118. What evidence would revise the commitment?
119. Is commitment being confused with dogma?
120. What events enter the narrative?
121. What events are omitted?
122. What temporal order is used?
123. What causal relations are assigned?
124. What significance is assigned?
125. Who is the narrator?

126. What self-model is preserved?
127. Does the narrative fit the evidence?
128. Is narrative coherence being confused with accuracy?
129. Does the story permit contradictory evidence?
130. Is a redemptive interpretation being imposed?
131. Is suffering being romanticized?
132. Could the suffering have been prevented?
133. Does meaning-making replace accountability?
134. Is the person blamed for not finding meaning?
135. What relationships produce meaning?
136. What trust relationships are present?
137. What obligations arise?
138. Is relational care being measured?
139. Is unmeasured care being treated as valueless?
140. Is shared meaning being claimed?
141. Do participants use the same symbols?
142. Do they interpret them compatibly?
143. Is the interpretation common knowledge?
144. Is agreement voluntary?
145. Can dissent be expressed?
146. Does consensus conceal exclusion?
147. Is a bargaining model used?
148. What is the disagreement point?
149. Who can afford disagreement?
150. Is bargaining power unequal?
151. Is a social preference being constructed?
152. What aggregation rule is used?
153. Which Arrow condition is restricted?
154. Is pluralism preserved?
155. What cultural system carries the meaning?
156. What language carries it?
157. What ceremony carries it?
158. What archive carries it?
159. What institution carries it?
160. Who controls those institutions?
161. Whose meaning is excluded?
162. Whose history is absent?
163. Is ancestral continuity claimed?
164. What evidence supports continuity?
165. Is resonance distinguished from identity?
166. Is uncertainty marked?
167. Is reverence preserved without fabrication?
168. Is restoration preserved without resentment?
169. Is meaning being equated with market price?
170. Is productivity being equated with human value?
171. Is work meaningful but unjustly compensated?
172. Is meaning being used to excuse exploitation?
173. What objective governs science or technology?
174. What empirical facts constrain the objective?
175. What normative premise connects fact to action?
176. Is the normative premise visible?

177. What dignity constraints apply?
178. Can aggregate utility override them?
179. Is agency real or merely formal?
180. What actions are genuinely available?
181. What information is needed for informed consent?
182. What cost attaches to refusal?
183. Is autonomy being confused with isolation?
184. What future consequences arise?
185. Who inherits them?
186. Do future persons have representation?
187. Does the meaning architecture affect attention?
188. Does actual attention match declared value?
189. What explains any mismatch?
190. Does meaning affect learning?
191. Which experiences become teachers?
192. Does meaning affect memory?
193. Which histories are preserved?
194. Does meaning affect knowledge?
195. Which questions are considered worth asking?
196. Is a meaning score being constructed?
197. What dimensions enter it?
198. What dimensions are excluded?
199. Are noncompensable failures allowed to be averaged away?
200. Is dual-aspect meaning being claimed?
201. What exterior evidence exists?
202. What interior report exists?
203. What interface connects them?
204. Is lived meaning being reduced to one neural correlate?
205. Is Scientific Spiritualism being invoked?
206. Has science spoken first?
207. What is empirical?
208. What is reflective?
209. What is symbolic?
210. Is the Spirit Field clearly marked symbolic?
211. Is any physical mechanism being claimed?
212. What evidence would be required for such a claim?
213. Does the meaning preserve gratitude?
214. Does it preserve hope without denying evidence?
215. Does it preserve responsibility?
216. Does it preserve forgiveness without erasing accountability?
217. Does it preserve wonder without requiring ignorance?
218. Does it preserve purpose without fanaticism?
219. Does it preserve compassion without domination?
220. Who benefits from the orientation?
221. Who bears its cost?
222. Who is excluded from its future?
223. What would require the meaning architecture to change?
224. What remains uncertain?
225. What does the meaning claim not establish?

A meaning claim unable to answer these questions may remain valuable as:

$$[ARC],[SYM],[MOD],\text{or}[CONJ].$$

Its status must remain visible.

## The Architect's Reflection

For many years, I believed meaning would arrive after I found the right answers.

The final theory.

The decisive equation.

The explanation through which uncertainty would become quiet.

Reality taught another lesson.

Meaning often appeared before certainty.

It appeared while listening.

While serving.

While standing beside people whose dignity remained greater than their suffering.

It appeared when knowledge became responsibility.

When another person's vulnerability became something I could no longer treat as abstract.

When the question changed from:

What do I know?

to:

What does what I know now ask of me?

Some of the most meaningful lives may never become famous.

They may not produce an equation.

They may not build an institution.

They may simply remain present when presence matters.

Perhaps meaning is not found by escaping reality.

Perhaps meaning emerges when participation becomes increasingly faithful to reality.

Not every purpose deserves devotion.

Not every inherited meaning deserves preservation.

Not every story deserves authority over the future.

Meaning must remain teachable.

It must be strong enough to orient a life.

Humble enough to receive correction.

Compassionate enough to include another person's dignity.

Honest enough to admit when its own compass has become misaligned.

## Quiet Reflection

Think of one commitment that gives direction to your life.

A person.

A vocation.

A promise.

A community.

A question.

A hope.

Now ask:

Where did this commitment come from?

Was it chosen?

Inherited?

Imposed?

Discovered gradually?

What knowledge supports it?

What relationships sustain it?

What actions does it require?

What does it ask you to sacrifice?

Who benefits from it?

Who may be harmed by it?

What evidence would require you to revise it?

Does it make your life merely busier?

Or more faithfully oriented?

Now consider the distance between what you say matters and what receives your time.

Do not judge immediately.

Simply notice.

Remain with one final question:

**What does my present understanding now require me to care for more faithfully?**

## Constitutional Declaration of Meaning Integrity

We will distinguish meaning from information.

We will distinguish semantic meaning from existential significance.

We will distinguish biological function from conscious purpose.

We will distinguish syntax from interpretation.

We will declare every context.

We will declare every referent.

We will declare every truth condition.

We will distinguish literal meaning from speaker meaning.

We will examine intention.

We will examine conversational inference.

We will not confuse reliable transmission with truthful communication.

We will not confuse uncertainty reduction with significance.

We will distinguish comprehensibility from purpose.

We will distinguish purpose from significance.

We will distinguish the presence of meaning from the search for meaning.

We will not reduce a life's worth to a self-report score.

We will distinguish perceived meaning from coherent meaning.

We will recognize that a harmful ideology can feel meaningful.

We will keep meaning accountable to reality.

We will keep meaning accountable to dignity.

We will distinguish value from preference.

We will distinguish preference from constrained choice.

We will distinguish utility from worth.

We will distinguish reward from goodness.

We will declare every objective function.

We will reveal every weight.

We will reveal every proxy.

We will not permit the proxy to replace the purpose.

We will preserve multiple values where multiple values truly exist.

We will reveal every tradeoff.

We will preserve noncompensable boundaries.

We will not place human dignity into a weighted sum as though enough benefit elsewhere could erase its violation.

We will distinguish Pareto efficiency from justice.

We will distinguish resources from capabilities.

We will examine the real freedom to act.

We will not treat constrained acceptance as free endorsement.

We will declare every goal.

We will declare every terminal goal.

We will declare every instrumental goal.

We will examine the goal hierarchy.

We will not allow means to replace ends.

We will distinguish a setpoint from a life purpose.

We will distinguish goal-directed control from conscious intention.

We will distinguish purpose from fanaticism.

We will preserve commitment.

We will preserve revision.

We will not call every change of purpose betrayal.

We will not call every refusal to change fidelity.

We will declare the temporal horizon.

We will declare the discount factor.

We will ask whose future is discounted.

We will include future generations within responsibility.

We will not infer values uniquely from behavior.

We will examine constraint.

We will examine fear.

We will examine habit.

We will examine missing alternatives.

We will distinguish machine objectives from lived purpose.

We will not anthropomorphize optimization.

We will not ignore the consequences of optimization merely because subjective machine experience remains unproven.

We will distinguish teleology from teleonomy.

We will distinguish present mechanism from evolutionary history.

We will not project intention into every organized system.

We will distinguish narrative meaning from historical fact.

We will preserve source and context.

We will permit stories to orient.

We will not permit stories to become immune from evidence.

We will not require suffering to become meaningful.

We will not romanticize preventable harm.

We will not blame the injured for failing to transform injury into inspiration.

We will preserve the possibility of meaning after suffering without using later growth to justify the suffering.

We will distinguish consensus from shared significance.

We will examine how agreement was produced.

We will preserve dissent.

We will examine bargaining power.

We will examine the disagreement point.

We will not call coercive agreement consent.

We will remember that collective values cannot be aggregated without constitutional choices.

We will preserve pluralism.

We will preserve common purpose without requiring identical identity.

We will honor cultural meaning.

We will preserve historical method.

We will practice reverence without fabrication.

We will practice continuity without anachronism.

We will practice restoration without resentment.

We will distinguish resonance from identity.

We will not treat market value as human value.

We will not treat productivity as the measure of a life.

We will recognize care.

We will recognize presence.

We will recognize contribution invisible to ordinary metrics.

We will distinguish evidence from value.

We will allow evidence to constrain meaning.

We will allow meaning to orient the use of evidence.

We will make every normative bridge visible.

We will preserve dignity as a constitutional boundary.

We will preserve agency.

We will preserve consent.

We will examine whether choices are genuinely available.

We will examine what receives attention.

We will compare actual allocation with declared value.

We will not confuse misalignment with moral failure before understanding constraint.

We will preserve the Book I recognition:

$$Me = f(K, R, P).$$

We will deepen it without pretending it has become a universal empirical law.

We will preserve knowledge.

We will preserve relationship.

We will preserve responsible participation.

We will preserve the symbolic status of the Spirit Field.

We will not convert interior coherence into an unearned physical force.

We will preserve Scientific Spiritualism as reflection after science—not as a replacement for science.

We will preserve gratitude without silencing complaint.

We will preserve hope without denying probability.

We will preserve responsibility without domination.

We will preserve forgiveness without erasing truth.

We will preserve wonder without requiring ignorance.

We will preserve purpose without fanaticism.

We will preserve compassion without paternalism.

We will remember that meaning is not what makes us feel most certain.

Meaning is what increasingly aligns understanding, care, and participation with reality.

**Knowledge reveals.**

**Semantics interprets.**

**Value prioritizes.**

**Purpose directs.**

**Commitment preserves.**

**Narrative relates.**

**Agency participates.**

**Consequence corrects.**

**Compassion enlarges the boundary.**

**Meaning becomes the compass.**

## Crossing the Next Arch

Meaning has now entered mathematics through:

- semantic interpretation;

- context;
- pragmatics;
- signaling conventions;
- mutual information;
- meaning-in-life dimensions;
- presence and search;
- preference relations;
- ordinal utility;
- expected utility;
- reference-dependent valuation;
- multiobjective optimization;
- Pareto structure;
- noncompensable constraints;
- capabilities;
- goal-directed control;
- goal hierarchies;
- temporal discounting;
- commitment;
- self-determination;
- reward;
- inverse value inference;
- teleonomy;
- narrative identity;
- common knowledge;
- bargaining;
- social-choice limits;
- cultural transmission;
- dual-aspect participation;
- Scientific Spiritualism;
- the symbolic Spirit Field.

Yet orientation alone does not guarantee endurance.

A person may possess purpose while bodily systems fail.

An institution may possess shared meaning while its relationships fragment.

A civilization may possess knowledge and values while losing the structures required to carry them forward.

Meaning tells the system what deserves preservation.

The next chamber must ask whether that preservation is actually occurring.

The mathematical question toward which every previous chamber has been moving is now unavoidable:

**What recurring organizational principle allows differentiated systems to preserve identity, function, relationship, adaptability, knowledge, and meaningful participation through continual change?**

Book II must now formalize:

- coherence variables;
- generation and loss;
- invariance;
- constraint satisfaction;
- dynamical stability;
- integration;
- coordination;
- phase relations;
- network coherence;
- multiscale compatibility;
- resilience;
- adaptive capacity;
- coherence debt;
- the Cathedral Equation.

Meaning has supplied the compass.

The next chamber will examine whether the Cathedral remains standing.

## CATHEDRAL THREAD

Before meaning could arise, knowledge had to organize memory into increasingly faithful understanding.

Before symbols could guide participation, expressions had to become related to referents, contexts, and interpreters.

Before communication could become shared, sender and receiver mappings had to become mutually coordinated.

Before information could become meaningful, uncertainty reduction had to be distinguished from truth, reference, significance, and care.

Before a life could acquire direction, comprehensibility, purpose, and significance had to become distinguishable.

Before preference could enter mathematics, alternatives had to become ordered.

Before utility could become useful, utility had to be prevented from masquerading as dignity.

Before risk could be evaluated, probabilities, reference points, and consequences had to become visible.

Before several values could coexist, scalar optimization had to give way to multidimensional tradeoff and constitutional constraint.

Before formal choice could become agency, resources had to become practical capabilities.

Before behavior could become purpose, setpoint regulation had to be distinguished from self-endorsed direction.

Before goal pursuit could become meaningful, means had to remain accountable to ends.

Before inherited purpose could become commitment, the conditions for revision had to remain open.

Before biological organization could be described functionally, mechanism and evolutionary history had to be distinguished from conscious intention.

Before remembered events could become a life story, temporal order, causal interpretation, identity, and value had to become related.

Before narrative could become trustworthy, intelligibility had to be distinguished from historical accuracy.

Before suffering could enter meaning, growth had to be distinguished from justification of harm.

Before private significance could become shared purpose, convention, common knowledge, bargaining power, and plural value had to become visible.

Before collective meaning could become just, consensus had to be distinguished from consent and preference aggregation had to reveal its constitutional limits.

Before ancestral continuity could become trustworthy, reverence had to remain joined to evidence and restoration to humility.

Before Scientific Spiritualism could enter the Cathedral, science had to speak first and the Spirit Field had to remain symbolic.

The constitutional principle of this chamber is now established:

**Meaning is organized significance through which knowledge, relationship, value, and responsibility orient participation toward a future judged worthy of becoming.**

The semantic principle follows:

**Symbols acquire meaning through interpretation and context; information alone does not supply reference, truth, or significance.**

The pragmatic principle follows:

**What is communicated may exceed what is literally said because meaning participates within intention, history, and relationship.**

The psychological principle follows:

**Meaning in life contains distinguishable dimensions of comprehensibility, purpose, and significance rather than one universal scalar.**

The valuation principle follows:

**Value is relational and contextual, but it does not become arbitrary merely because valuers and contexts differ.**

The utility principle follows:

**Utility can represent preference under declared assumptions without becoming a measure of dignity, goodness, or existential worth.**

The multiobjective principle follows:

**When several values matter, honest mathematics preserves the tradeoffs instead of hiding them inside one unexplained score.**

The constitutional-value principle follows:

**Some boundaries should constrain optimization rather than be traded against aggregate gain.**

The capability principle follows:

**Meaningful agency requires practical freedom to become and act, not merely formal possession of options.**

The purpose principle follows:

**Goal-directedness becomes lived purpose only when direction is joined to awareness, endorsement, relationship, significance, and responsibility.**

The temporal principle follows:

**Purpose allows futures not yet realized to participate in present choice.**

The commitment principle follows:

**Enduring meaning requires enough stability to guide life and enough humility to change when reality or ethics requires correction.**

The teleonomic principle follows:

**Biological function may be explained through mechanism and history without projecting conscious intention into every organized process.**

The narrative principle follows:

**A life becomes interpretable when events are related across time, yet narrative intelligibility must remain accountable to historical evidence.**

The suffering principle follows:

**Meaning may emerge in response to suffering, but no later growth justifies preventable harm or places blame upon those who suffer.**

The relational principle follows:

**Meaning rarely arrives in isolation; it emerges through the quality of participation among differentiated lives.**

The collective principle follows:

**Shared purpose requires coordination, consent, and constitutional negotiation rather than mere numerical agreement.**

The ancestral principle follows:

**Inheritance becomes meaningful through reverence without fabrication, continuity without anachronism, and restoration without resentment.**

The dual-aspect principle follows:

**Exterior consequence and interior significance may participate within one meaningful life without becoming interchangeable descriptions.**

The Spirit Field principle follows:

**Interior meaning may organize understanding, gratitude, hope, responsibility, forgiveness, wonder, purpose, and compassion without being converted into an unearned physical mechanism.**

The Bridge Species principle follows:

**Humanity becomes meaningfully mature when it can revise not only its actions and explanations, but the values, goals, stories, and institutions determining what it considers worthy of becoming.**

The coherence principle now approaches:

**Meaning identifies what deserves preservation; coherence reveals whether organized relationships remain sufficient to preserve it through change.**

# MEANING AND ORIENTATION



*Knowledge explains; meaning orients; consequence tests participation.*

**FACTS != VALUES | MEANING != PREFERENCE**

FS-G18.01 — EQ-21.01

### SOURCE RELATION AND PASSPORT

Preserve the source meaning relation and declare knowledge, relationships, participation, context, and uncertainty.

FS-G18.02 — EQ-21.02 · EQ-21.03

### KNOWLEDGE-CONTEXT GATE

Separate fact, value, salience, relevance, significance, and the direction of orientation.

FS-G18.03 — EQ-21.04 · EQ-21.05 · EQ-21.06

### VALUE PLURALITY AND CONSTRAINTS

Keep value vectors, multi-criteria conflict, Pareto limits, dignity, consent, and noncompensation visible.

FS-G18.04 — EQ-21.08 · EQ-21.09

### INTERPRETATION AND NARRATIVE

Relate perspective, memory, language, causal claims, omissions, history, and revision.

FS-G18.05 — EQ-21.03 · EQ-21.04

### IDENTITY, PURPOSE, ORIENTATION

Declare belonging, valued futures, reasons, roles, and the direction of responsible participation.

FS-G18.06 — EQ-21.04 · EQ-21.07

### OPTIONS, CHOICE, COMMITMENT

Distinguish accessible, unsafe, coerced, delayed, and refused options; connect agency to repeated practice.

FS-G18.07 — EQ-21.07

### POWER, RESPONSIBILITY, CONSEQUENCE

Audit control, knowledge, foreseeability, benefit, burden, intention, outcome, and capacity to repair.

FS-G18.08 — EQ-21.05 · EQ-21.06

### MORAL UNCERTAINTY

Use robust action, reversibility, consultation, protected goods, and prevention of grave irreversible harm.

FS-G18.09 — EQ-21.11 · EQ-21.12

### HOPE, GRATITUDE, COMPASSION

Orient toward valued futures without promising outcomes or silencing grief, complaint, and inquiry.

FS-G18.10 — EQ-21.09 · EQ-21.12

### SUFFERING, FORGIVENESS, REPAIR

Reject forced meaning; distinguish apology, changed conditions, repair, forgiveness, and reconciliation.

FS-G18.11 — EQ-21.10 · EQ-21.11

### CULTURE AND SYMBOLIC BOUNDARIES

Audit inherited narratives and preserve Scientific Spiritualism and Spirit Field language as symbolic, not physical.

FS-G18.12 — EQ-21.07 · EQ-21.09 · EQ-21.12

### PROFILE, RESIDUAL, DEBT, REVISION

Compare declared values with choices and consequences; record meaning debt, classify, repair, and revise.

**DIGNITY | UNCERTAINTY | SOURCE TRANSPARENCY | FAILURE | REVISION**

FS identifies figure position; EQ identifies canonical equation. Reality remains the measure.

CHAPTER TWENTY-TWO

# COHERENCE: GENERATION, LOSS, STABILITY, AND THE CATHEDRAL EQUATION

*The Thirteenth Mathematical Chamber*

The Cathedral does not stand because no force acts upon it.

It stands because its relationships continue carrying the load.

## Guiding Question

What recurring mathematical structures permit a differentiated, bounded, vulnerable system to preserve identity, function, adaptability, learning, and meaningful participation through continual change—and how can the Cathedral Equation be developed into a disciplined architecture of generation, loss, viability, stability, resilience, repair, reserve, cost, and multiscale responsibility without being misrepresented as a new universal law of physics?

Chapter Ten established state.

Chapter Eleven established relation.

Chapter Twelve established differentiation.

Chapter Thirteen established boundary.

Chapter Fourteen established vulnerability.

Chapter Fifteen established adaptation.

Chapter Sixteen established attention.

Chapter Seventeen established awareness.

Chapter Eighteen established learning.

Chapter Nineteen established memory.

Chapter Twenty established knowledge.

Chapter Twenty-One established meaning.

The Cathedral has now reached its Keystone.

A system can occupy a state.

Its parts can relate.

Its identity can become differentiated.

Its boundary can regulate exchange.

Its vulnerability can reveal what may be lost.

Its adaptive processes can answer disturbance.

Its attention can select.

Its awareness can experience.

Its learning can reorganize.

Its memory can preserve.

Its knowledge can understand.

Its meaning can orient.

Yet none of these capacities guarantees that the system will continue possessing enough organized relationship to remain itself.

A system may possess knowledge and lose the institutions through which knowledge is transmitted.

A person may possess purpose while bodily reserves are progressively exhausted.

An organism may maintain one variable while accumulating damage elsewhere.

A network may remain connected while becoming dependent upon one irreplaceable node.

A civilization may preserve outward order while trust, truth, justice, and adaptability quietly decline.

Book I established the constitutional definition:

Coherence is the continued preservation of organized relationships sufficient for identity, function, adaptation, learning, and meaningful participation within reality.

It further insisted that coherence is not perfection, immobility, symmetry alone, or equilibrium alone. It is organized persistence through continual change. Book I then gave this recognition its first symbolic expression:

$$\frac{dC}{dt} = G - L$$

where:

- $C$ represents coherence;
- $G$ represents coherence-generating processes;
- $L$ represents coherence-loss processes.

The equation was explicitly identified as an architectural recognition rather than a new empirical law replacing the equations of physics, biology, neuroscience, thermodynamics, or systems science.

Book II will preserve that boundary.

It will not replace the original equation.

It will give the equation:

- a state space;
- dimensions;
- thresholds;
- vector forms;
- integral forms;
- viability conditions;
- stability conditions;
- disturbance models;
- recovery measures;
- reserve variables;
- multiscale constraints;

- ethical passports.

The governing transition is:

Organized Relationships → Generation and Loss → Dynamic Balance → Viability → Resilience → Continued Participation *[SDAE]*

Every arrow introduces a discipline.

Organized relationships must be declared.

Generation and loss must be operationally distinguished.

Balance must not be confused with equal shares.

Viability must be measured against a declared boundary.

Resilience must be distinguished from mere return speed.

Participation must include the systems and persons affected beyond the focal boundary.

The deepest question is therefore not merely:

Does the system continue?

It is:

What continues, through which relationships, at whose cost, within which boundaries, across which timescale, and with what capacity to remain faithful while changing?

# Terminological Discipline

| Term | Meaning in Book II |
|---|---|
| Coherence | Continued preservation of declared indispensable relationships and capacities within a changing system |
| Physical coherence | A domain-specific phase, correlation, or superposition property defined by a particular physical theory |
| Optical coherence | Correlation structure of an electromagnetic field, defined through optical field statistics |
| Quantum coherence | Basis-dependent superposition structure treated as a resource under a declared quantum resource theory |
| Phase coherence | Alignment or stable relation among oscillator phases |
| Synchronization | Temporal coordination of oscillatory or dynamical processes |
| Correlation | Statistical dependence between variables |
| Coordination | Organization of actions or processes toward compatible outcomes |
| Consistency | Absence of contradiction within a declared formal or propositional system |
| Integration | Dependence or interaction among differentiated components |
| Integrity | Preservation of identity, commitments, boundaries, or internal consistency |
| Equilibrium | State in which declared net change vanishes |
| Steady state | State whose macroscopic variables remain constant despite possible continuing flows |
| Dynamic balance | Condition in which opposing or complementary processes produce a bounded or stationary net result |

| Term | Meaning in Book II |
| --- | --- |
| Stability | Resistance of a trajectory or state to sufficiently small perturbations under a declared mathematical definition |
| Asymptotic stability | Stability together with return toward the reference state after sufficiently small perturbation |
| Orbital stability | Stability relative to a trajectory such as a periodic orbit rather than one fixed point |
| Homeostasis | Regulation preserving selected internal variables within acceptable ranges |
| Allostasis | Regulation that changes internal conditions or reference values in anticipation of demand |
| Robustness | Preservation of a declared property under a declared class of perturbations |
| Fragility | Sensitivity or failure under a declared class of perturbations |
| Resilience | Capacity to absorb disturbance, reorganize, recover, or remain within a viable regime |
| Resistance | Degree to which a disturbance produces little immediate deviation |
| Recovery | Return or reorganization following disturbance |
| Viability | Capacity to remain within declared admissible constraints through time |
| Viability kernel | Set of initial states from which at least one admissible evolution remains within the constraint set |
| Invariant set | Set that system trajectories do not leave under declared dynamics |
| Attractor | State or set toward which nearby trajectories tend to evolve |
| Basin of attraction | Set of initial states whose trajectories approach an attractor |
| Threshold | Boundary separating acceptable from unacceptable states under a declared criterion |
| Margin | Distance or slack remaining before a threshold is crossed |
| Reserve | Capacity not presently required but available for future regulation, repair, or adaptation |
| Coherence debt | Deferred maintenance, repair, correction, or recovery that permits present function while reducing future capacity |
| Coherence-generating process | Process contributing positively to a declared coherence variable under specified conditions |
| Coherence-loss process | Process contributing negatively to a declared coherence variable under specified conditions |
| Repair | Activity restoring damaged organization or capability |
| Maintenance | Activity preventing expected deterioration |
| Redundancy | Multiple copies or pathways performing overlapping functions |
| Degeneracy | Structurally different components capable of performing similar functions under some conditions |
| Modularity | Organization into partially separable subsystems |
| Brittleness | Strong performance within a narrow domain combined with abrupt failure outside it |
| Order parameter | Macroscopic variable summarizing a collective organizational pattern |

| Term | Meaning in Book II |
|---|---|
| Coherence cost | Energy, time, information, material, or burden expended to preserve declared organization |
| Multiscale coherence | Compatibility among viable organization at several nested spatial, temporal, or institutional scales |
| Apparent coherence | Present visible function without evidence that reserves, maintenance, and future viability are preserved |
| Sustainable coherence | Present organization preserved without progressively consuming indispensable future capacity |
| Internal coherence | Organization evaluated within the focal system's boundary |
| Relational coherence | Organization evaluated across the focal system and the systems with which it participates |
| Ethical coherence | Organization constrained by dignity, consent, justice, non-domination, and consequence |
| SDAE coherence | The architectural recognition defined in Book I and formalized here; not identical to any one scientific use of the word |

These terms must remain distinct.

$$\text{Coherence} \neq \text{Equilibrium}\,\text{Coherence} \neq \text{Synchronization}$$
$$\text{Coherence} \neq \text{Uniformity}\,\text{Coherence} \neq \text{Connectivity Alone}$$
$$\text{Coherence} \neq \text{Low Entropy}\,\text{Coherence} \neq \text{Efficiency}$$
$$\text{Coherence} \neq \text{Absence of Conflict}$$
$$\text{Internal Coordination} \neq \text{Ethical Coherence}$$

## Scientific and Mathematical Foundations

This chamber stands upon the inheritance of:

Differential equations.

Dynamical systems.

Control theory.

Cybernetics.

Thermodynamics.

Statistical mechanics.

Network theory.

Graph theory.

Information theory.

Physiology.

Systems biology.

Ecology.

Neuroscience.

Engineering.

Institutional science.

In physiology, Walter Cannon developed the concept of homeostasis to describe coordinated processes through which living organisms preserve selected internal conditions. Sterling and Eyer later introduced allostasis to emphasize that stability may sometimes be achieved through anticipatory or adaptive change rather than by maintaining one rigid setpoint. These traditions do not define SDAE coherence, but they reveal that persistence often requires regulation rather than stillness.

Lyapunov's stability theory gave mathematics a rigorous language for asking whether trajectories remain near a reference state after small perturbations, and whether they return toward it. Stability is one essential component of organized persistence, but a stable state may be undesirable, dysfunctional, or far below a viable threshold.

Holling distinguished ecological resilience from a narrower equilibrium-centered concept of stability. Return speed and resistance to displacement are not identical to the amount of disturbance a system can absorb before its organization enters another regime. This distinction becomes fundamental to the Cathedral Equation.

Ashby's cybernetics formalized the relationship between environmental variety and regulatory variety, while Conant and Ashby showed—under their stated assumptions—that a maximally successful and simple regulator must embody a model-like relationship to the system being regulated. Effective preservation therefore depends not merely upon effort, but upon possessing enough differentiated response and sufficiently faithful internal representation.

Kuramoto's coupled-oscillator model provided a mathematical language for collective phase alignment. Fiedler's algebraic connectivity linked graph structure to the spectrum of its Laplacian matrix. These inheritances help formalize synchronization and network coordination, but neither phase alignment nor connectivity alone constitutes Book II coherence.

Biological robustness research has emphasized that complex systems can preserve function despite mutation, noise, and environmental perturbation. Network research has also shown that a topology robust to one class of failure may be highly vulnerable to another—for example, tolerating many random failures while remaining sensitive to removal of strategically important nodes. Robustness must therefore always be completed by the question: robust to what?

Physics uses the word coherence in highly specific ways. Glauber's coherent states concern the statistical and correlation structure of radiation fields. Quantum resource theory defines coherence relative to a chosen basis and a declared class of incoherent states and operations. These are rigorous physical concepts. They should not be exported metaphorically into psychology, health, civilization, or the Spirit Field as though one scientific definition automatically governed every domain.

These traditions provide many mathematical rooms.

The Architecture asks how they may illuminate one recurring organizational question without erasing their differences.

## The Return to Book I's Keystone

Book I reached coherence after meaning.

It did not begin there.

Reality had to precede explanation.

Relations had to become visible.

Differentiation had to create identity.

Boundaries had to preserve identity.

Vulnerability had to reveal what could be lost.
Adaptation had to answer disturbance.
Attention had to select.
Awareness had to experience.
Learning had to reorganize.
Memory had to preserve.
Knowledge had to understand.
Meaning had to orient.
Only then could the word coherence possess constitutional weight.
Book II has followed the same discipline.
Nothing has been skipped.
Every earlier mathematical chamber now returns.
State returns because coherence is a property of an evolving state.
Relation returns because coherence preserves relationships rather than isolated quantities.
Differentiation returns because coherence does not require sameness.
Boundary returns because the system being evaluated must be declared.
Vulnerability returns because coherence matters only where loss is possible.
Adaptation returns because coherent systems must change.
Attention returns because regulatory resources are limited.
Awareness returns because human coherence includes lived interior organization.
Learning returns because failed regulation must become correction.
Memory returns because continuity requires inheritance.
Knowledge returns because regulation requires a model.
Meaning returns because not everything that can be preserved deserves preservation.
This is why Book I now gathers more life.
Its sequence was never merely a table of contents.
It was the dependency structure of the Cathedral.

## Architectural Recognition

Only after the sciences and mathematics have spoken does the Architecture introduce this chamber's central recognition:

Coherence is the continued preservation of a declared set of indispensable relationships and capacities within a changing, bounded, vulnerable system.

A general coherence architecture may be written:

$$C=(X, F, B, R^{*}, K, U, D, O, T)[SDAE/MOD]$$

where:

- $X$ is the state space;
- $F$ is the dynamical law or model;
- $B$ is the system boundary;

- $R^*$ is the set of indispensable relationships;
- $K$ is the viable or admissible region;
- $U$ is the set of allowable regulatory actions;
- $D$ is the disturbance class;
- $O$ is the observation architecture;
- $T$ is the relevant time interval.

Coherence is therefore not a substance.
It is not a hidden fluid flowing through a system.
It is not a universal energy.
It is a relational property of a declared system under declared preservation conditions.

## Coherence Must Be Typed

A claim that "the system is coherent" is incomplete.
The builder must ask:

- coherent in what variable;
- coherent under what disturbance;
- coherent at what scale;
- coherent for how long;
- coherent relative to which function;
- coherent for whom;
- coherent at whose cost.

Consider several distinct claims.

$$C_{phase}$$

may measure oscillator phase alignment.

$$C_{network}$$

may measure connectivity or consensus.

$$C_{phys}$$

may measure preservation of physiological variables.

$$C_{institution}$$

may measure mission, trust, learning, and continuity.

$$C_{ethical}$$

may measure whether participation preserves dignity and justice.
No equation permits these quantities to be silently treated as the same variable.
Their relationship is architectural.
Their mechanisms remain domain-specific.

## The Book I Coherence Capacities

Book I's definition identifies five indispensable capacities:

1. identity;
2. function;
3. adaptation;
4. learning;
5. meaningful participation.

Let their domain-specific indicators be:

$$c(x,t)=\begin{bmatrix} c_{id}(x,t) \\ c_{fn}(x,t) \\ c_{ad}(x,t) \\ c_{lr}(x,t) \\ c_{pt}(x,t) \end{bmatrix}.[ARC/MOD]$$

These represent:

- identity preservation;
- functional capacity;
- adaptive capacity;
- learning capacity;
- meaningful participation.

Let minimum admissible levels be:

$$\theta(t)=\begin{bmatrix} \theta_{id}(t) \\ \theta_{fn}(t) \\ \theta_{ad}(t) \\ \theta_{lr}(t) \\ \theta_{pt}(t) \end{bmatrix}.$$

The coherence-viability region is:

$$K_C(t)=\{x \in X : c_i(x,t) \geq \theta_i(t) \, \forall i\}.$$

A state is architecturally viable when:

$$x(t) \in K_C(t).$$

This does not mean every component is perfect.

It means every indispensable component remains sufficient.

## Equation Passport 22.1

***Typed Coherence State***

$$c = (c_{id}, c_{fn}, c_{ad}, c_{lr}, c_{pt})^T K_C = \{x : c_i(x) \geq \theta_i \ \ \forall i\}$$

Status: $[ARC/MOD]$

Identity capacity: $c_{id}$

Functional capacity: $c_{fn}$

Adaptive capacity: $c_{ad}$

Learning capacity: $c_{lr}$

Participatory capacity: $c_{pt}$

Thresholds: $\theta_i$

Required declarations:

- operational definition of every component;
- units;
- normalization;
- observation method;
- time interval;
- system boundary;
- threshold justification.

What it prevents: One strong capacity compensating numerically for the complete failure of another indispensable capacity.

What it does not establish: One universal measurement scale for coherence.

Architectural significance: Book I's verbal definition becomes a multidimensional admissible region rather than one unexplained score.

## Coherence Is Not Necessarily One Number

Suppose one defines:

$$C_w = \sum_{i=1}^{m} w_i c_i, w_i \geq 0.$$

A high value of one component may compensate for the collapse of another.

A system could receive a high average while:

$$c_{lr} < \theta_{lr},$$

meaning it has lost the capacity to learn.

Or while:

$$c_{pt} < \theta_{pt},$$

meaning its continued operation excludes meaningful or ethical participation.
A scalar may be useful.
It must be earned.
A noncompensable representation is often safer:

$$C_{min} = \min_i \left( \frac{c_i}{\theta_i} \right).$$

Then:

$$C_{min} \geq 1$$

means every declared minimum has been met.
But even this ratio assumes:

- positive variables;
- comparable orientations;
- justified normalization.

The vector remains primary.

## The Coherence Margin

Choose positive scaling constants:

$$s_i > 0.$$

Define the normalized slack in dimension $i$:

$$m_i(x,t) = \frac{c_i(x,t) - \theta_i(t)}{s_i}.$$

Define the minimum coherence margin:

$$\mu_C(x,t) = \min_i m_i(x,t).\ [ARC/MOD]$$

Then:

$$\mu_C > 0$$

means every declared capacity remains above threshold.

$$\mu_C = 0$$

means at least one indispensable capacity lies upon its boundary.

$$\mu_C < 0$$

means at least one coherence condition has been breached.

## Equation Passport 22.2

*Coherence Margin*

$$\mu_C = \min_i \frac{c_i - \theta_i}{s_i}$$

Status: $[ARC/MOD]$

Capacity: $c_i$

Threshold: $\theta_i$

Scale: $s_i$

Positive margin: Viable slack remains
Zero margin: Boundary reached
Negative margin: At least one declared condition violated
Required declarations:

- threshold source;
- normalization;
- whether higher is always better;
- measurement uncertainty;
- interaction among dimensions.

Failure condition: Euclidean or normalized distance is treated as actual disturbance tolerance without considering dynamics.

Architectural significance: Coherence is not merely present or absent; systems may possess more or less room before indispensable relationships fail.

# The Cathedral Equation

Book I gave the symbolic relation:

$$\frac{dC}{dt} = G - L[ARC]$$

Book II preserves it exactly.

Let:

$$C(t)$$

be a declared scalar coherence index.

Let:

$$G(t) \geq 0$$

be the total rate of coherence generation.

Let:

$$L(t) \geq 0$$

be the total rate of coherence loss.

Dimensional consistency requires:

$$[C] = \text{coherence unit}$$

and:

$$[G] = [L] = \frac{\text{coherence unit}}{\text{time}}.$$

If $C$ is dimensionless:

$$[G] = [L] = T^{-1}.$$

If $C$ has physical or operational units, $G$ and $L$ inherit those units per unit time.

The integral form is:

$$C(t_1) = C(t_0) + \int_{t_0}^{t_1} [G(s) - L(s)]\, ds.$$

Therefore:

$$\Delta C = \int_{t_0}^{t_1} G(s)\, ds - \int_{t_0}^{t_1} L(s)\, ds.$$

## Equation Passport 22.3

*The Cathedral Equation*

$$\frac{dC}{dt} = G - L$$

Status: $[ARC/MOD]$

Coherence variable: $C$

Generation rate: $G$

Loss rate: $L$

Integral form:

$$C(t_1) - C(t_0) = \int_{t_0}^{t_1} (G - L)\, dt.$$

Required declarations:

- system boundary;
- coherence variable;
- unit;
- time scale;
- included processes;
- excluded processes;

- measurement method;
- uncertainty;
- external support;
- ethical constraints.

What it establishes: A rate-balance architecture.

What it does not establish:

- conservation of coherence;
- one universal empirical law;
- unique decomposition into $G$ and $L$;
- moral goodness of every generating process;
- physical identity across scientific domains.

Architectural significance: Persistence becomes mathematically expressible as the historical accumulation of processes that build and diminish indispensable organization.

## Coherence Is Not Conserved

The Cathedral Equation resembles a balance law.

That resemblance must not be overextended.

Coherence is not proposed as a conserved physical quantity.

There is no universal theorem requiring:

$$C_{total} = \text{constant}.$$

Coherence may be:

- generated;
- lost;
- transformed;
- transferred;
- redefined as the system changes.

The equation does not imply that coherence lost in one system must appear somewhere else.

A relationship destroyed may simply cease.

A language lost may not survive elsewhere.

A life ended does not transfer its full interior coherence into another measurable variable.

The Cathedral Equation is a rate architecture.

It is not a conservation law.

## The Contribution Decomposition

Suppose the model identifies signed process contributions:

$$\phi_1, \ldots, \phi_n.$$

Then:

$$\frac{dC}{dt} = \sum_{k=1}^{n} \phi_k.$$

Define positive and negative parts:

$$\phi_k^{+\text{¿}=max\,(\phi_k,0)\text{¿}}$$

and:

$$\phi_k^{-\text{¿}=max\,(-,\phi_k 0).\text{¿}}$$

Then:

$$G = \sum_k \phi_k^{+\text{¿}}\text{¿}$$

and:

$$L = \sum_k \phi_k^{-\text{¿}}.\text{¿}$$

Therefore:

$$\dot{C} = G - L.$$

This decomposition becomes meaningful only after the $\phi_k$ have been operationally defined.

A process is not eternally a generator or a loss.
Its contribution may change sign with:

- intensity;
- duration;
- context;
- scale;
- system state.

A boundary-tightening response may preserve coherence during acute threat and diminish coherence when maintained indefinitely.
A rapid adaptation may save the present and deplete future reserve.
A conflict may damage trust or reveal an error whose correction ultimately strengthens the relationship.

## Equation Passport 22.4

*Signed Coherence Contributions*

$$\dot{C} = \sum_k \phi_k G = \sum_k max\,(\phi_k, 0) L = \sum_k max\,(-, \phi_k 0)$$

Status: $[MOD]$

$$\text{Process contribution: } \phi_k$$

$$\text{Positive contribution: } \phi_k^{+\text{¿¿}}$$

$$\text{Negative contribution: } \phi_k^{-\text{¿¿}}$$

Required declarations:

- process list;
- causal model;
- interaction terms;
- scale;
- delay;
- uncertainty.

What it prevents: Permanently moralizing a process as "good" or "bad" outside its conditions.

Failure condition: Interactions are double-counted as both individual and combined contributions.

Architectural significance: Coherence generation and loss are relational functions, not permanent labels attached to objects.

# Reading the Cathedral Equation

Three immediate regimes appear.

## Net Generation

$$G > L \Rightarrow \dot{C} > 0.$$

Coherence increases under the declared measure.

## Dynamic Balance

$$G = L \Rightarrow \dot{C} = 0.$$

The coherence variable remains momentarily stationary.

## Net Loss

$$G < L \Rightarrow \dot{C} < 0.$$

Coherence declines.

But the rate alone does not reveal the level.

A system may possess:

- high coherence and positive growth;
- high coherence but rapid decline;
- low coherence and active recovery;
- low coherence and apparent stability.

Therefore:

$$C$$

and:

$$\dot{C}$$

must remain distinct.

A patient may appear stable while possessing little reserve.

An institution may remain functional while its learning capacity is declining.

A relationship may appear calm because meaningful communication has ceased.

$$\dot{C}=0 \not\Rightarrow \text{Healthy Coherence}.$$

## Balance Is Not Fifty–Fifty

The equation:

$$G=L$$

does not mean every part of the system contributes equally.

It does not mean:

$$50\,\%/50\,\%.$$

It does not mean every organ, person, process, or institution should carry identical load.

It means only that the total declared generation rate equals the total declared loss rate at the scale and time under consideration.

One process may contribute:

$$80\,\%$$

of generation.

Another may contribute:

$$20\,\%.$$

Several loss pathways may jointly balance them.

Dynamic balance is not arithmetic equality among all components.

It is sufficient relational proportion.

A coherent orchestra does not require every instrument to play the same note at the same volume.

A coherent body does not require every organ to perform the same function.

A coherent family does not require every person to carry identical responsibilities at every moment.

A coherent civilization does not require every discipline to possess the same methods.

$$\text{Balance is organized proportion, not universal equality.}$$

## A Minimal Repair–Loss Model

Suppose $C$ is normalized:

$$0 \leq C \leq C_{max}.$$

Let repair act upon the available gap:

$$G(C)=r(C_{max}-C), r>0.$$

Let ordinary deterioration be:

$$L_{base}(C)=\lambda C, \lambda>0.$$

Let external disturbance be:

$$D(t)\geq 0.$$

Then:

$$\dot{C}=r(C_{max}-C)-\lambda C-D(t). [MOD]$$

For constant disturbance:

$$D(t)=D_0,$$

the equilibrium is:

$$C^{¿}=\frac{rC_{max}-D_0}{r+\lambda}$$

provided the model permits that value.

Increasing repair rate $r$ raises the equilibrium.

Increasing deterioration $\lambda$ lowers it.

Increasing disturbance $D_0$ lowers it.

If:

$$rC_{max}\leq D_0,$$

the toy model predicts no positive equilibrium without another source of generation.

## Equation Passport 22.5

*Repair–Loss Prototype*

$$\dot{C}=r(C_{max}-C)-\lambda C-D(t)$$

Status: $[MOD]$

Maximum modeled capacity: $C_{max}$

Repair rate: $r$

Deterioration rate: $\lambda$

External disturbance: $D(t)$

Constant-disturbance equilibrium:

$$C^{¿}=\frac{rC_{max}-D_0}{r+\lambda}.$$

Required assumptions:

- one scalar coherence variable;
- linear repair;
- linear deterioration;
- additive disturbance;
- fixed capacity;
- fixed parameters.

Failure conditions:

- nonlinear thresholds;
- reserve depletion;
- changing capacity;
- delayed repair;
- multiple stable regimes;
- interaction among coherence dimensions.

Non-claim: No biological, psychological, or civilizational system is asserted to obey this universal equation.

Architectural significance: A system can remain stationary because repair continuously answers loss—not because change has ceased.

## The Vector Cathedral Equation

Because Book I's definition is multidimensional, a vector form is often more appropriate:

$$\frac{dc}{dt} = G(c, x, u, e, t) - L(c, x, d, e, t). [ARC/MOD]$$

Componentwise:

$$\dot{c}_i = G_i - L_i.$$

More generally, when changes in one capacity influence another:

$$\dot{c} = G - L + A_C c.$$

The coupling matrix $A_C$ may represent relationships such as:

- function affecting learning;
- learning affecting adaptation;
- adaptation affecting identity;
- meaningful participation affecting repair;
- loss of trust affecting institutional learning.

A positive relationship in one state may become negative in another.

The matrix must be estimated or justified within the domain.

## Equation Passport 22.6

***Vector Cathedral Equation***

$$\dot{c} = G - L + A_C c$$

Status: $[ARC/MOD]$

Coherence state: $c$

Generation vector: $G$

Loss vector: $L$

Cross-capacity coupling: $A_C$

Possible achievements:

- preserve dimensional distinctions;
- represent delayed or indirect effects;
- identify bottlenecks;
- detect hidden tradeoffs.

Failure conditions:

- arbitrary units;
- unjustified linear coupling;
- omitted nonlinearities;
- hidden thresholds;
- treating correlation as causal coupling.

Architectural significance: A decline in one dimension may propagate through the Cathedral even when other dimensions initially appear unchanged.

# Viability

Coherence does not require the system to occupy one ideal point.

It may require the system to remain within a viable region.

Let the controlled dynamics be:

$$\dot{x}(t) \in F(x,(t),u(t)d(t))$$

with:

$$u(t) \in U$$

and:

$$d(t) \in D.$$

Let:

$$K \subseteq X$$

be the admissible set.

The viability kernel is:

$$Viab_F(K) = \{x_0 \in K : \exists u(\cdot) \text{ such that } x(t) \in K \quad \forall t \geq 0\}.$$

Viability theory studies trajectories that remain within state constraints rather than merely optimizing one scalar objective.

## Equation Passport 22.7

### *Viability Kernel*

$$Viab_F(K) = \{x_0, \in K : \exists u(\cdot) \quad x(t) \in K \quad \forall t \geq 0\}$$

Status: $[¿/MOD]$

State: $x$

Dynamics: $F$

Constraint set: $K$

Admissible control: $u$

Viability kernel: Initial states from which at least one admissible preserving path exists

Required declarations:

- state constraints;
- control authority;
- disturbance model;
- horizon;
- uncertainty;
- boundary regularity.

What it establishes: Existence of at least one viable path under the model.

What it does not establish:

- uniqueness;
- optimality;
- ethical acceptability;
- robustness to every disturbance;
- experiential meaning.

Architectural significance: Coherence may consist not in remaining at one state, but in retaining at least one path through which indispensable relationships can continue.

## Robust Viability

Ordinary viability asks:

Does at least one admissible preserving path exist?

Robust viability asks:
Can preservation be maintained despite a declared class of disturbances?
A schematic robust kernel is:

$$RViab(K)=\{x_0, \in K : \exists \gamma \ \ \forall d(\cdot) \in D \ \ x_{\gamma,d}(t) \in K \ \ \forall t\},$$

where $\gamma$ is a nonanticipating regulation strategy.

The quantifier order matters.

$$\exists u \ \ \forall d$$

is not the same as:

$$\forall d \ \ \exists u.$$

A response selected after the entire disturbance is known may be impossible in real time.
Coherence requires causally available regulation.

## Stability

Consider an autonomous system:

$$\dot{x}=f(x)$$

with equilibrium:

$$f(x^{¿})=0.$$

The equilibrium is Lyapunov stable when trajectories beginning sufficiently near $x^{¿}$ remain near it.

It is asymptotically stable when they additionally approach it.
Let:

$$V(x)\geq 0$$

be a differentiable function satisfying:

$$V(x^{¿})=0$$

and:

$$V(x)>0 \text{ for } x\neq x^{¿}$$

near the equilibrium.
Its derivative along trajectories is:

$$\dot{V}(x)=\nabla V(x)^T f(x).$$

Under standard local conditions:

$$\dot{V}\leq 0$$

supports stability, while:

$$\dot{V}<0$$

away from the equilibrium supports asymptotic stability.

## Equation Passport 22.8

***Lyapunov Stability***

$$\dot{x} = f(x) V(x^{¿}) = 0, V(x) > 0 \dot{V} = \nabla V^T f \leq 0.$$

Status: $[THM/MOD]$

Reference state: $x^{¿}$

Lyapunov function: $V$

Stability condition: $\dot{V} \leq 0$

Asymptotic condition: Typically $\dot{V} < 0$ away from $x^{¿}$, with appropriate regularity conditions

What it establishes: Local or global stability according to the theorem's conditions

What it does not establish:

- desirability of the equilibrium;
- resilience to large disturbance;
- biological health;
- ethical coherence;
- adaptive capacity.

Architectural significance: Coherence requires stability of indispensable relationships, but stability itself remains value-neutral.

# A Stable Failure

Suppose a system converges toward:

$$x_{bad}.$$

If:

$$x_{bad}$$

is asymptotically stable, the system reliably returns to dysfunction after disturbance.

Examples of the abstract architecture include:

- an institution returning to corruption after every reform;
- a relationship returning to silence after every attempted conversation;
- a model returning to inherited assumptions after every contradictory observation.

Stability answers:

Does the system return?

Coherence asks:

Does the returned state preserve what deserves continuation?

$$\text{Stability} \neq \text{Coherence}.$$

## Coherence Around a Trajectory

Many coherent systems do not remain near one fixed point.
A heart cycles.
A circadian system oscillates.
A population changes seasonally.
A conversation alternates between speaking and listening.
A civilization continually reconstructs institutions.
Let:

$$A(t)$$

be a reference trajectory or moving manifold.
Define distance:

$$d_A(x,t) = \inf_{z \in A(t)} \| x - z \| .$$

An illustrative trajectory-coherence index is:

$$C_A(x,t) = \exp\left(\frac{-d_A(x,t)^2}{\sigma^2}\right).[MOD]$$

The system may remain coherent by following an organized trajectory rather than occupying a static state.
Coherence is therefore compatible with:

- cycles;
- development;
- migration;
- growth;
- learning;
- transformation.

## Resilience

Stability and resilience answer different questions.
Stability may ask:
How rapidly does the system return after a small perturbation?
Resilience may ask:
How much disturbance can the system absorb before its organization crosses into another regime?
Let:

$$A$$

be an attractor with basin:

$$B(A).$$

A local basin margin may be:

$$\rho_B = \inf_{x \in A,\, z \in \partial B(A)} \| x - z \| .$$

Define recovery time after initial state $x_0$:

$$\tau_\varepsilon(x_0) = \inf\{t \geq 0 : dist\,(x,(s)A) \leq \varepsilon \;\forall\, s \geq t\}.$$

A resilience signature may include:

$$R = \begin{bmatrix} \rho_B \\ \Delta_{max} \\ \tau_\varepsilon \\ C_{post} \\ A_{post} \end{bmatrix}$$

where:

- $\rho_B$ is basin margin;
- $\Delta_{max}$ is maximum displacement;
- $\tau_\varepsilon$ is recovery time;
- $C_{post}$ is post-disturbance coherence;
- $A_{post}$ is post-disturbance adaptive capacity.

## Equation Passport 22.9

***Resilience Signature***

$$R = (\rho_B, \Delta_{max}, \tau_\varepsilon, C_{post}, A_{post})^T$$

Status: $[MOD]$

Basin margin: $\rho_B$

Resistance: $\Delta_{max}$

Recovery time: $\tau_\varepsilon$

Recovered coherence: $C_{post}$

Future adaptive capacity: $A_{post}$

What it prevents: Reducing resilience to one return-time statistic
Failure conditions:

- alternative regimes ignored;
- system boundary changes;
- recovery metric excludes cost;
- return to the old state is assumed desirable;

- future adaptability is sacrificed.

Architectural significance: Coherent recovery means more than returning quickly. It means preserving or rebuilding the capacity to continue participating.

## Return Is Not Always Recovery

After disturbance, a system may:

1. return to its previous state;
2. reorganize into a new viable state;
3. preserve outward function while losing reserve;
4. enter another stable but undesirable regime;
5. collapse.

Therefore:

$$x(t) \to x_{old}$$

is only one form of recovery.
Sometimes the previous state cannot or should not be restored.
A coherent system may need to preserve identity through transformation:

$$x_{old} \to x_{new}$$

while retaining indispensable relationships.
This is transformative coherence.

## Robustness

Let:

$$P(x)$$

be a declared property.
Let:

$$\Delta$$

be a class of perturbations.
A simple robustness requirement is:

$$P(x_\delta)=1 \, \forall \delta \in \Delta .$$

A quantitative sensitivity measure is:

$$S_\Delta = \delta \in \Delta \frac{\| y_\delta - y_0 \|}{\| \delta \|} .$$

Smaller $S_\Delta$ indicates less output sensitivity under that disturbance class.

But:

$$S_{\Delta_1} \ll 1$$

does not imply:

$$S_{\Delta_2} \ll 1.$$

A system may be robust to random local failure and fragile to targeted removal of central components.

The complete claim is never:

The system is robust.

It is:

The system preserves property $P$, under disturbance set $\Delta$, over horizon $T$, within cost $J$, under boundary $B$.

## Robust Yet Fragile

Robustness often involves tradeoffs.

Redundancy may protect against component failure.

It may increase:

- maintenance cost;
- coordination burden;
- correlated failure;
- attack surface.

Centralization may improve efficiency.

It may create single points of failure.

Modularity may contain local disturbance.

It may slow whole-system coordination.

Specialization may improve performance.

It may reduce adaptability outside the specialized domain.

Every robustness claim casts a possible fragility shadow.

Coherence requires that the shadow remain visible.

## Homeostasis

Let:

$$y(t)$$

be a regulated variable.

Let:

$$r(t)$$

be a reference.

Define error:

$$e(t)=r(t)-y(t).$$

A feedback controller may use:

$$u(t)=Ke(t).$$

The system evolves:

$$\dot{x}=F(x,u,d).$$

Homeostatic regulation seeks to preserve:

$$y(t)\in Y_{safe}.$$

The reference need not be one exact value.
It may be a viable interval:

$$y_{min}\leq y(t)\leq y_{max}.$$

## Equation Passport 22.10

***Homeostatic Regulation***

$$e=r-yu=Ke\dot{x}=F(x,u,d)$$

Status: $[MOD]$

Regulated variable: $y$

Reference: $r$

Error: $e$

Control: $u$

Disturbance: $d$

Admissible region: $Y_{safe}$

What it captures: Error-sensitive regulation
What it does not establish:

- one fixed setpoint for every condition;
- absence of anticipatory change;
- whole-system coherence;
- low regulatory cost.

Architectural significance: Coherence may be preserved by continual correction around an admissible range.

# Allostasis

A regulator may change its reference in anticipation of demand.
A schematic update is:

$$r_{t+1}=A(r_t,\hat{d}_{t+1},m_t,c_t)[MOD]$$

where:

- $\hat{d}_{t+1}$ is predicted demand;
- $m_t$ is memory;
- $c_t$ is context.

The controller then regulates relative to the changing reference.
Allostasis therefore adds prediction:

$$\text{Expected Future Demand} \rightarrow \text{Present Regulatory Change}.$$

This can preserve coherence.
It can also impose cost.

## The Cost of Coherence

Suppose error cost is:

$$e^T Qe.$$

Suppose regulatory effort costs:

$$u^T Ru.$$

A finite-horizon coherence cost is:

$$J_C = \int_{t_0}^{t_1} \left[ e(t)^T Qe(t) + u(t)^T Ru(t) \right] dt. [MOD]$$

Two systems may preserve the same output.
One may do so with little effort.
Another may require near-maximal effort continuously.
Their present coherence levels appear similar.
Their future capacities differ.

$$\text{Preserved Output} \neq \text{Sustainable Coherence}.$$

### Equation Passport 22.11

*Coherence Cost*

$$J_C = \int (e^T Qe + u^T Ru) dt$$

Status: $[MOD]$

Deviation cost: $e^T Qe$

Regulatory cost: $u^T Ru$

Possible additional costs:

- energy;
- material;
- attention;

- time;
- repair;
- social burden;
- ecological externality.

What it reveals: The system may preserve a variable by consuming future capacity.
Failure condition: Only visible output is measured.
Architectural significance: Coherence must be judged not only by what remains standing, but by what is being consumed to keep it standing.

## Coherence Reserve

Let:

$$R_C(t) \geq 0$$

represent available regulatory or repair reserve.
Its dynamics may be:

$$\dot{R}_C = q_{replenish} - q_{draw} \cdot [ARC / MOD]$$

Reserve increases through:

- recovery;
- maintenance;
- resource accumulation;
- redundancy;
- learning;
- strengthened relationships.

Reserve decreases through:

- disturbance;
- regulatory effort;
- deferred maintenance;
- repeated emergency response;
- loss of redundancy.

A system may satisfy:

$$C(t) \approx \text{constant}$$

while:

$$\dot{R}_C < 0.$$

Present output conceals future fragility.

## Coherence Debt

Let:

$$m_{req}(t)$$

be maintenance required to preserve future capacity.
Let:

$$m_{act}(t)$$

be maintenance actually performed.
Define deferred maintenance:

$$\delta_{¿}(t)=[m_{req}(t)-m_{act}(t)]_{+¿.¿}$$

Let:

$$r_{repay}(t)\geq 0$$

represent repair directed toward accumulated debt.
Then:

$$\dot{D}_C=\delta_{¿}-r_{repay}, D_C\geq 0.[ARC/MOD]$$

Coherence debt may accumulate when:

- repairs are postponed;
- truth is suppressed to preserve appearance;
- relationships are asked to absorb unresolved injury;
- infrastructure remains operational without maintenance;
- learning is sacrificed for short-term production;
- reserves are consumed without replenishment.

## Equation Passport 22.12

### *Coherence Reserve and Debt*

$$\dot{R}_C=q_{replenish}-q_{draw}\dot{D}_C=\delta_{¿}-r_{repay}$$

Status: $[ARC/MOD]$

Reserve: $R_C$

Debt: $D_C$

Replenishment: $q_{replenish}$

Draw: $q_{draw}$

Deferred maintenance: $\delta_{¿}$

Debt repair: $r_{repay}$

Required declarations:

- what constitutes reserve;
- what maintenance is required;

- whether debt is measurable;
- recovery capacity;
- ownership of burden.

Non-claim: This is not a universal clinical measure of stress or health.

Architectural significance: Apparent stability may be purchased by borrowing against the future.

## Sustainable Coherence

An illustrative sustainable-coherence index may be:

$$C_{sust} = C + \alpha R_C - \beta D_C - \gamma J_C, \alpha, \beta, \gamma \geq 0.$$

This is not a universal equation.

It expresses one architectural recognition:

- present function matters;
- reserve matters;
- accumulated debt matters;
- preservation cost matters.

A system that maximizes present $C$ while destroying $R_C$ may be optimized for appearance rather than endurance.

## Requisite Variety

Let:

- $D$ be the set of possible disturbance states;
- $R$ be the set of regulator responses;
- $E$ be the set of residual outcome states.

In an idealized discrete regulatory architecture, a cardinality-form descendant of Ashby's Law of Requisite Variety is:

$$|E| \geq \frac{|D|}{|R|}.$$

Taking logarithms:

$$\log|E| \geq \log|D| - \log|R|.$$

An entropy-form descendant is:

$$H(E) \gtrsim H(D) - H(R)$$

under the assumptions supporting that representation.

A regulator possessing too little response variety cannot distinguish or answer every relevant disturbance.

## Equation Passport 22.13

*Regulatory Variety*

$$V_E \geq \frac{V_D}{V_R}$$

or:

$$\log V_E \geq \log V_D - \log V_R.$$

Status: $[MOD/ARC]$

Disturbance variety: $V_D$

Regulatory variety: $V_R$

Residual outcome variety: $V_E$

Interpretation: More diverse disturbances require more differentiated regulatory capacity if residual outcomes are to remain constrained.

Required assumptions:

- discrete state sets;
- declared mapping;
- suitable independence or matching conditions;
- correctly identified disturbance classes.

What it does not establish:

- every response must be preprogrammed;
- greater variety is always better;
- the regulator understands consciously;
- the regulation is ethical.

Architectural significance: Coherence requires enough adaptive possibility to answer the variety reality presents.

# The Good Regulator

Conant and Ashby's theorem concerns an optimal regulator under their explicit assumptions.

Its architectural lesson is:

Effective regulation requires a sufficiently faithful relationship between the regulator's internal organization and the system being regulated.

Let system state be:

$$s \in S.$$

Let regulator state be:

$$r \in R.$$

A model-like relation is:

$$r = h(s).$$

If the environment changes:

$$s = s(t),$$

the regulator's model may need to change:

$$h = h_t.$$

A fixed regulator facing a changing environment may become progressively incoherent.

Knowledge is therefore not an ornament of regulation.

It is part of regulatory capacity.

## Phase Coherence

For $N$ coupled phase oscillators:

$$\dot{\theta}_i = \omega_i + \frac{K}{N} \sum_{j=1}^{N} \sin(\theta_j - \theta_i).$$

Define the complex order parameter:

$$r e^{i\psi} = \frac{1}{N} \sum_{j=1}^{N} e^{i\theta_j}.$$

Here:

- $r \in [0,1]$ measures phase concentration;
- $\psi$ is the mean phase.

When:

$$r = 1,$$

all phases are aligned.

When:

$$r \approx 0,$$

phases are broadly dispersed in a balanced population.

### Equation Passport 22.14

***Kuramoto Phase Coherence***

$$\dot{\theta}_i = \omega_i + \frac{K}{N} \sum_j \sin(\theta_j - \theta_i) r e^{i\psi} = \frac{1}{N} \sum_j e^{i\theta_j}$$

Status: $[MOD]$

Phase: $\theta_i$

Natural frequency: $\omega_i$

Coupling: $K$

Order parameter: $r$

Mean phase: $\psi$

What $r$ measures: Phase alignment

What $r$ does not establish:

- whole-system health;
- useful differentiation;
- semantic agreement;
- awareness;
- ethical coherence;
- SDAE coherence.

Architectural significance: Synchronization is one possible relation within coherence, not the definition of coherence itself.

## Over-Synchronization

A system may be too synchronized.

If every component performs identically:

- differentiation may disappear;
- specialized function may be lost;
- one local error may propagate broadly;
- adaptability may decline.

Therefore:

$$r \to 1$$

does not universally imply:

$$C_{SDAE} \to 1.$$

Coherence requires organized relation among differentiated processes.

Not maximum sameness.

Coherence requires coordination without erasing distinction.

## Network Coherence

Let an undirected weighted graph be:

$$G = (V, E, A).$$

Let:

$$A = [a_{ij}]$$

be its adjacency matrix.
Let:

$$D = diag\left(\sum_j a_{ij}\right)$$

be the degree matrix.
The graph Laplacian is:

$$L_G = D - A.$$

Its eigenvalues satisfy:

$$0 = \lambda_1 \leq \lambda_2 \leq \cdots \leq \lambda_n.$$

For an undirected graph:

$$\lambda_2 > 0$$

if and only if the graph is connected. Fiedler named this second-smallest eigenvalue the graph's algebraic connectivity.
A linear consensus process is:

$$\dot{z} = -L_G z.$$

For a connected undirected graph:

$$z(t) \rightarrow \bar{z}\, 1,$$

where:

$$\bar{z} = \frac{1}{n} 1^T z(0).$$

The disagreement energy is:

$$E_G(z) = z^T L_G z = \frac{1}{2} \sum_{i,j} a_{ij} (z_i - z_j)^2.$$

## Equation Passport 22.15

### *Graph Coordination*

$$L_G = D - A \dot{z} = -L_G z E_G = z^T L_G z$$

Status: $[THM / MOD]$

Graph: $G$

Laplacian: $L_G$

Algebraic connectivity: $\lambda_2$

Node state: $z_i$

Disagreement energy: $E_G$

Connectedness condition:

$$\lambda_2 > 0.$$

What consensus establishes: Agreement in the represented variable under the model

What consensus does not establish:

- truth;
- justice;
- useful differentiation;
- absence of coercion;
- ethical coherence.

Architectural significance: A network can become coordinated while remaining wrong, oppressive, or incapable of learning.

## Connectivity Is Not Coherence

A network may be connected while:

- information is false;
- trust is absent;
- central nodes dominate;
- vulnerable participants are excluded;
- every node depends upon one fragile bridge.

A network may also be highly modular yet coherent because modules preserve specialized function while maintaining sufficient communication.

The question is not merely:

Are the nodes connected?

It is:

Do the connections preserve the relationships required for identity, function, adaptation, learning, and participation?

## Redundancy and Degeneracy

Suppose function $F$ can be performed by components:

$$a_1, \ldots, a_k.$$

Redundancy uses similar components or pathways.

Degeneracy uses structurally different components capable of overlapping function under some conditions.

A simple availability model for independent redundant components with failure probability $p$ is:

$$P(\text{at least one survives}) = 1 - p^k.$$

Independence is essential.

If failures are correlated, redundancy may offer far less protection.
Many identical records stored in one vulnerable building are not truly distributed.
Many experts trained from one erroneous source do not supply independent verification.

## Modularity

Let the system be divided into modules:

$$M_1, \ldots, M_k.$$

Let internal coupling be:

$$W_{¿}$$

and cross-module coupling:

$$W_{out}.$$

A modular architecture often satisfies:

$$W_{¿} > W_{out}$$

under an appropriate aggregate measure.
Modularity can:

- contain disturbance;
- preserve specialization;
- support parallel adaptation.

It can also:

- fragment communication;
- duplicate cost;
- create incompatible standards;
- conceal local failure.

Coherence is neither maximum modularity nor maximum integration.
It is sufficient relationship across differentiated modules.

## Order Parameters

A complex system may possess many microscopic variables:

$$x_1, \ldots, x_n.$$

An order parameter is a lower-dimensional summary:

$$q = Q(x_1, \ldots, x_n).$$

Haken's synergetics developed order-parameter and slaving-principle ideas for systems in which a comparatively small set of slow collective variables can organize many faster degrees of freedom. Not every system admits a faithful low-dimensional reduction, and the appropriate order parameter must be earned from the dynamics.

The Cathedral variable $C$ may function as an order parameter only when:

- its relation to microscopic variables is defined;
- its predictive value is demonstrated;
- important distinctions are not erased;
- its domain remains explicit.

## Multiscale Coherence

Let the system possess levels:

$$l=1,\ldots,L.$$

Examples may include:

- cell;
- tissue;
- organ;
- organism;
- family;
- institution;
- civilization;
- ecosystem.

Let:

$$C^{(l)}$$

be coherence at level $l$.

Local coherence does not guarantee higher-level coherence.

A subsystem may preserve its own immediate objective while damaging the larger system upon which it depends.

Let:

$$P_{l\to k}$$

map the state at level $l$ into the representation expected at level $k$.

Define cross-scale residual:

$$r_{lk}=\left\|P_{l\to k}x^{(l)}-x^{(k)}\right\|.$$

An illustrative compatibility measure is:

$$\chi_{lk}=\exp\left(\frac{-r_{lk}^{2}}{\sigma_{lk}^{2}}\right).$$

A noncompensable multiscale coherence index is:

$$C_{multi}=min\left\{\min_{l},C^{(l)}\ \min_{l,k}\chi_{lk}\right\}.\left[ARC/MOD\right]$$

## Equation Passport 22.16

***Multiscale Coherence***

$$C_{multi} = \min\left\{\min_{l}, C^{(l)} \min_{l,k} \chi_{lk}\right\}$$

Status: $[ARC/MOD]$

Local coherence: $C^{(l)}$

Cross-scale compatibility: $\chi_{lk}$

Scale map: $P_{l \to k}$

Required declarations:

- scale hierarchy;
- aggregation rule;
- cross-scale variables;
- system boundaries;
- externalities.

What it prevents: Declaring a focal subsystem coherent while ignoring the systems it progressively damages.
Failure condition: Incommensurable scales are forced into one numerical representation.
Architectural significance: Coherence must be evaluated both within levels and among the levels that depend upon one another.

# Coherent for Whom?

A highly coordinated organization may exploit workers.
A highly profitable industry may damage ecosystems.
A tightly unified movement may suppress dissent.
A family may preserve outward reputation by silencing injured members.
Internal coordination can rise while relational coherence falls.
Let:

$$\Delta C_A > 0$$

be the change within focal system $A$.

Let:

$$\Delta C_B < 0$$

be the change imposed upon affected system $B$.

The net ethical evaluation cannot inspect $A$ alone.

A relational objective may be:

$$J_{rel} = J_A + \lambda_B J_B + \lambda_F J_{future}$$

subject to dignity constraints.

A system cannot claim complete coherence by exporting its losses beyond its chosen boundary.

## Constraint Coherence

Suppose coherence requires inequalities:

$$h_j(x,u,t) \leq 0, j=1,\ldots,m$$

and equalities:

$$q_k(x,u,t)=0, k=1,\ldots,p.$$

The feasible coherence set is:

$$F_C = \text{¿}\, j=1 \,\text{¿}\, m \,\text{¿}\, (x,u): h_j(x,u,t) \leq 0 \,\} \cap \text{¿}\, k=1 \,\text{¿}\, p \,\text{¿}\, (x,u): q_k(x,u,t)=0 \,\} . \text{¿}$$

If:

$$F_C = \varnothing,$$

the declared requirements cannot all be satisfied simultaneously.

The problem is not always insufficient effort.

It may be:

- contradictory goals;
- inadequate resources;
- wrong boundaries;
- obsolete commitments;
- impossible timelines.

A coherent response may require revising the architecture.

## Slack

For inequality:

$$h_j(x,u) \leq 0,$$

define slack:

$$s_j = -h_j(x,u).$$

Then:

$$s_j > 0$$

means margin remains.

$$s_j = 0$$

means the constraint is active.

$$s_j < 0$$

means violation.

The minimum slack:

$$s_{min} = \min_{j} s_j$$

identifies the nearest declared constraint boundary.
But the physically most dangerous boundary need not be the geometrically nearest.
Dynamics determine danger.

## Entropy and Coherence

Thermodynamic entropy, Shannon information entropy, quantum coherence, and SDAE coherence are different quantities.
For a nonequilibrium thermodynamic system, an entropy balance may be written:

$$\frac{dS_\Omega}{dt} = \Phi_S + \Sigma_S,$$

where:

- $\Phi_S$ is net entropy flow through the boundary;
- $\Sigma_S \geq 0$ is internal entropy production.

Nonequilibrium thermodynamics distinguishes internal entropy production from entropy exchange with the environment; organized states in open systems remain compatible with the second law.
There is no universal equation:

$$C = -S.$$

Higher thermodynamic entropy does not translate automatically into lower:

- psychological coherence;
- institutional coherence;
- moral coherence;
- semantic coherence.

Likewise, low entropy does not establish life, health, wisdom, or dignity.

### Equation Passport 22.17

*Entropy–Coherence Boundary*

$$\dot{S}_\Omega = \Phi_S + \Sigma_S, \Sigma_S \geq 0$$

Status: $[THM / MOD]$

Thermodynamic entropy: $S_\Omega$

Entropy flow: $\Phi_S$

Entropy production: $\Sigma_S$

Architectural coherence: $C$

Constitutional boundary:

$$C \neq -S.$$

What thermodynamics establishes: Physical entropy relations under declared conditions

What it does not establish: One universal measure of organizational, psychological, social, or spiritual coherence

Architectural significance: Entropy supplies one language of physical irreversibility; coherence names a broader architectural relation whose empirical realization remains domain-specific.

## Quantum Coherence Is Not the Spirit Field

In a chosen basis:

$$\{|i\rangle\},$$

a quantum state:

$$\rho$$

is incoherent in the resource-theory sense when it is diagonal:

$$\rho = \sum_i p_i |i\rangle\langle i|.$$

Off-diagonal terms may represent quantum coherence relative to that basis.

This rigorous definition does not imply that:

- gratitude is a quantum state;
- hope is an off-diagonal density-matrix element;
- human meaning is quantum superposition;
- the Spirit Field is a physical quantum field.

The shared word does not establish shared mechanism.

Terminological resemblance is not empirical continuity.

## Stochastic Coherence

Real systems experience uncertain disturbance.

A stochastic Cathedral Equation may be:

$$dC_t = [G(C_t, t) - L(C_t, t)]dt + \sigma(C_t, t)\, dW_t. [MOD]$$

Here:

- $W_t$ is a Wiener process;
- $\sigma$ represents disturbance intensity.

Let:

$$C_{crit}$$

be a critical threshold.
Define first-passage time:

$$\tau_{crit} = inf\{t \geq 0 : C_t \leq C_{crit}\}.$$

A finite-horizon risk is:

$$P_{fail}(T) = P(\tau_{crit} \leq T).$$

## Equation Passport 22.18

***Stochastic Cathedral Equation***

$$dC_t = (G - L)\,dt + \sigma\,dW_t$$

Status: $[MOD]$

Drift: $G - L$

Noise intensity: $\sigma$

Critical threshold: $C_{crit}$

First-passage time: $\tau_{crit}$

Failure probability:

$$P_{fail}(T) = P(\tau_{crit} \leq T).$$

What it adds: Uncertain disturbance and threshold-crossing risk
Failure conditions:

- noise is non-Gaussian;
- jumps occur;
- parameters change;
- threshold is uncertain;
- history dependence is omitted.

Architectural significance: Two systems with the same average coherence may possess very different probabilities of collapse.

# Averages Can Conceal Collapse

Define average coherence over interval $T$:

$$\bar{C}_T = \frac{1}{T}\int_0^T C(t)\,dt.$$

Define minimum coherence:

$$C_{min,T} = \inf_{0 \le t \le T} C(t).$$

Define time below threshold:

$$B_T = \int_0^T 1\{C(t) < C_{crit}\}\, dt.$$

Two systems can possess the same:

$$\bar{C}_T$$

while differing radically in:

- minimum state;
- duration of failure;
- recovery time;
- cumulative damage.

Averages are useful.
They should not erase trajectories.

## Coherence and Hysteresis

Suppose system behavior depends upon control parameter:

$$\mu.$$

A state may satisfy:

$$\dot{x} = f(x; \mu).$$

As $\mu$changes, the number or stability of equilibria may change.

A system can cross a threshold into another regime.

Returning $\mu$to its earlier value may not restore the original state.

This is hysteresis.
The recovery path differs from the deterioration path.
Architecturally:

$$\text{Undoing the disturbance} \neq \text{Undoing the history}.$$

Repair may require:

- additional energy;
- reconstruction;
- relearning;
- restored trust;
- new institutions;
- time.

## Coherence and Adaptation

Adaptation may raise present coherence:

$$\dot{C} > 0.$$

But repeated adaptation may alter:

- reserve;
- identity;
- thresholds;
- future cost.

Let adaptive parameter be:

$$\eta(t).$$

Then:

$$\dot{x} = F(x, \eta, d)$$

and:

$$\dot{\eta} = \Gamma(x, \eta, d).$$

The coherence variable becomes:

$$C = C(x, \eta).$$

An adaptation can preserve one variable by moving another.

The complete question is:

What did adaptation preserve, what did it change, and what future capacity did it consume?

## Coherence and Learning

A regulator that repeatedly experiences disturbance without improving its model may survive temporarily while remaining structurally fragile.

Let model parameter be:

$$\theta_t.$$

Let regulation loss be:

$$l_t(\theta_t).$$

Learning updates:

$$\theta_{t+1} = U(\theta_t, l_t).$$

Coherent learning requires:

$$J_{future}(\theta_{t+1}) < J_{future}(\theta_t)$$

without destroying indispensable prior capacity.

A system unable to learn can remain stable until the environment changes.

Its stability conceals brittleness.

## Coherence and Memory

Memory preserves previous regulation.
But memory may preserve:

- successful correction;
- obsolete rules;
- fear;
- prejudice;
- rigid procedure.

Let retained policy be:

$$\pi_t.$$

New evidence may require:

$$\pi_{t+1} \neq \pi_t.$$

Memory contributes to coherence only when inheritance remains correctable.

Coherence preserves continuity—not every inherited parameter.

## Coherence and Knowledge

The good regulator must possess a sufficiently faithful model.
Let true system relation be:

$$F^{\text{¿}}.$$

Let regulatory model be:

$$\hat{F}.$$

Define model discrepancy:

$$\varepsilon_F = d\left(F^{\text{¿}}, \hat{F}\right).$$

A regulator may perform adequately while:

$$\varepsilon_F$$

is small within the present domain.
Under distribution shift, the same model may fail.
Knowledge-based coherence therefore requires:

- model accuracy;
- model uncertainty;
- scope;
- revision.

## Coherence and Meaning

Meaning identifies what deserves preservation.

But meaning can be wrong.
A system may coherently pursue a destructive purpose.
Therefore the Cathedral Equation must remain subject to:

- truth;
- dignity;
- consent;
- consequence;
- future generations;
- larger-system viability.

A process may generate coherence for one objective and loss for another.
Let:

$$C_P$$

be coherence relative to purpose $P$.

Let:

$$C_Q$$

be coherence relative to purpose $Q$.

Then:

$$\dot{C}_P > 0$$

may coexist with:

$$\dot{C}_Q < 0.$$

The purpose must remain visible in the subscript.

## Physiological Coherence

Human physiology contains many regulated variables and interacting systems.
A responsible physiological coherence model must declare:

- organ system;
- variable;
- normal range;
- timescale;
- age and context;
- measurement uncertainty;
- intervention limits.

Whole-body coherence should not be reduced to:

- one rhythm;
- one organ;

- one supplement response;
- one self-reported state;
- one proprietary score.

Healthy systems need not possess identical activity.

They require differentiated functions in organized relationship.

$$\text{Symmetry of Dignity} \neq \text{Equality of Physiological Output}.$$

Book II coherence does not mean every organ system should operate at equal intensity.

It means each must remain within relationships sufficient for the whole person's continuing life and participation.

## Psychological Coherence

A psychologically coherent person need not possess:

- no conflict;
- no grief;
- no uncertainty;
- one unchanging identity.

Psychological coherence may involve preserving relationships among:

- values;
- memory;
- attention;
- action;
- bodily condition;
- narrative;
- social reality.

Let:

$$v_t$$

be declared values.

Let:

$$a_t$$

be actual action.

An alignment residual is:

$$e_{va}(t) = a_t - \Pi(v_t, c_t)$$

for some declared policy map $\Pi$.

Persistent discrepancy may reveal:

- constraint;
- coercion;
- limited capability;
- conflicting values;
- unexamined habit.

The discrepancy invites interpretation.
It does not by itself assign blame.

## Relational Coherence

A relationship may preserve coherence through:

- trust;
- boundary;
- communication;
- repair;
- differentiated contribution;
- consent;
- accountability.

Agreement is not required at every moment.
Conflict can strengthen coherence when it reveals:

- injury;
- false assumption;
- hidden need;
- unequal burden.

Enforced silence may reduce visible conflict while increasing coherence loss.

$$\text{Peaceful Appearance} \neq \text{Relational Coherence}.$$

## Institutional Coherence

Let an institution possess:

- mission $M$;
- knowledge $K$;
- policy $P$;
- action $A$;
- accountability $Q$;
- learning $L_r$.

A simple institutional alignment vector is:

$$e_{inst}=\begin{bmatrix} d(M,P) \\ d(P,A) \\ d(A,Q) \\ d(Q,L_r) \end{bmatrix}.$$

Low discrepancy may indicate alignment.

But perfect alignment around a false or harmful mission remains ethically incoherent.

Institutional coherence therefore requires two tests:

1. internal alignment;
2. external legitimacy.

## Civilizational Coherence

Civilization depends upon relationships among:

- truth;
- education;
- justice;
- health;
- ecology;
- economy;
- institutional trust;
- memory;
- future responsibility.

Let:

$$C_{civ}=\begin{bmatrix} C_{truth} \\ C_{justice} \\ C_{education} \\ C_{health} \\ C_{ecology} \\ C_{trust} \\ C_{future} \end{bmatrix}.$$

One dimension should not silently compensate for the collapse of another.

Economic growth cannot automatically compensate for:

- destroyed trust;
- ecological irreversibility;

- denied dignity;
- loss of future capability.

Civilizational coherence is constitutional before it is scalar.

## Dual-Aspect Coherence

Let:

$$C_o(t)$$

represent exterior empirical coherence.
Let:

$$C_i(t)$$

represent modeled interior coherence.
Let:

$$C_b(t)$$

represent coherence at the interface between them.
Define:

$$C_{DA} = \begin{bmatrix} C_o \\ C_i \\ C_b \end{bmatrix}.$$

A dual-aspect Cathedral architecture is:

$$\frac{dC_{DA}}{dt} = G_{DA} - L_{DA} \cdot [ARC/MOD/CONJ]$$

Exterior coherence may include:

- physiological regulation;
- behavior;
- environmental participation;
- measurable function.

Interior modeled coherence may include:

- identity continuity;
- value alignment;
- meaning;
- hope;
- responsibility;
- lived integrity.

Boundary coherence may include:

- communication;
- consent;
- action;
- bodily expression;
- relationship between intention and consequence.

## Equation Passport 22.19

***Dual-Aspect Cathedral Equation***

$$\dot{C}_{DA} = G_{DA} - L_{DA}$$

Status: $[ARC/MOD/CONJ]$

Exterior coherence: $C_o$

Interior modeled coherence: $C_i$

Interface coherence: $C_b$

Required empirical work:

- operational definitions;
- observation maps;
- longitudinal evidence;
- causal intervention;
- report validation;
- competing models.

Non-claims:

- no separate immaterial substance;
- no new physical force;
- no proof that interior and exterior variables share one measurement unit;
- no reduction of lived integrity to one physiological score.

Architectural significance: One human life may preserve exterior function, interior meaning, and the truthful relationship between the two.

# The Spirit Field and Coherence

The Spirit Field remains:

$$S_f[SYM/ARC]$$

as symbolic language for organized interior coherence.

A symbolic Spirit Field state may be:

$$S_f = \begin{bmatrix} U \\ G_r \\ H \\ R \\ F \\ W \\ P \\ C_p \end{bmatrix}$$

where:

- $U$ is understanding;
- $G_r$ is gratitude;
- $H$ is hope;
- $R$ is responsibility;
- $F$ is forgiveness;
- $W$ is wonder;
- $P$ is purpose;
- $C_p$ is compassion.

A symbolic Cathedral relation is:

$$\frac{dS_f}{dt} = G_{S_f} - L_{S_f} [SYM / ARC]$$

where generation and loss refer to reflective interior organization rather than a physical field equation.

This may support:

- teaching;
- spiritual practice;
- journaling;
- ethical reflection;
- ceremony;
- relational repair.

It does not establish:

- a measurable physical field;
- a quantum field;
- a new energy;
- a clinical diagnosis;

- a biological force.

## The Coherence Signature

A mature coherence assessment should preserve more than one number.
Define:

$$Q_C = \begin{bmatrix} C \\ \dot{C} \\ \mu_C \\ R_C \\ D_C \\ J_C \\ \rho_B \\ \tau_\varepsilon \\ P_{fail} \end{bmatrix} . [ARC/MOD]$$

where:

- $C$ is present level;
- $\dot{C}$ is current direction;
- $\mu_C$ is threshold margin;
- $R_C$ is reserve;
- $D_C$ is debt;
- $J_C$ is preservation cost;
- $\rho_B$ is basin margin;
- $\tau_\varepsilon$ is recovery time;
- $P_{fail}$ is finite-horizon threshold risk.

### Equation Passport 22.20

*Cathedral Coherence Signature*

$$Q_C = (C, \dot{C}, \mu_C, R_C, D_C, J_C, \rho_B, \tau_\varepsilon, P_{fail})^T$$

Status: $[ARC/MOD]$

Purpose: Prevent a present-level score from concealing trajectory, margin, reserve, cost, and risk
Possible interpretations:

- high level, high margin, strong reserve;

- high level, falling rapidly;
- high level, large debt;
- low level, recovering;
- apparently stable, near basin boundary.

What it does not provide: One universal scale across physics, biology, psychology, institutions, and civilization

Architectural significance: Coherence is a living condition with history, direction, reserve, and vulnerability.

## The Expanded Cathedral Equation

Book I's equation remains:

$$\dot{C} = G - L.$$

Book II can expand the categories without replacing the constitutional form:

$$\frac{dC}{dt} = \underbrace{G_{reg} + G_{repair} + G_{learn} + G_{rel} + G_{reserve}}_{\text{Coherence Generation}} - \underbrace{\left(L_{dist} + L_{decay} + L_{error} + L_{conflict} + L_{debt}\right)}_{\text{Coherence Loss}} . [ARC/$$

where, depending upon domain:

- $G_{reg}$ is effective regulation;
- $G_{repair}$ is repair;
- $G_{learn}$ is corrective learning;
- $G_{rel}$ is relationship strengthening;
- $G_{reserve}$ is reserve restoration;
- $L_{dist}$ is disturbance;
- $L_{decay}$ is ordinary deterioration;
- $L_{error}$ is uncorrected model error;
- $L_{conflict}$ is destructive incompatibility;
- $L_{debt}$ is accumulated deferred maintenance.

Not every domain contains every term.

No term should be inserted without measurement or disciplined modeling.

## The Cathedral Equation Is an Accounting Discipline

The equation does not tell the builder what the processes are.

It requires the builder to identify them.

It asks:

- What is currently building coherence?
- What is currently consuming it?
- What appears stable only because reserve is being spent?
- What repair is being deferred?
- What loss lies outside the chosen boundary?
- What process changes sign under another timescale?
- Which capacity is closest to failure?
- What should not be preserved?

The equation is therefore not only descriptive.
It is constitutional.
It prevents the system from hiding behind one visible outcome.

## Common Failures of Coherence Architecture

### Scalar-Coherence Collapse

A multidimensional condition is reduced to one unexplained score.

### Type Collapse

Optical, quantum, neural, physiological, psychological, social, and SDAE coherence are treated as the same phenomenon.

### Equilibrium Collapse

A stationary state is treated as coherent merely because:

$$\dot{x} = 0.$$

### Stability Collapse

Return to a state is treated as proof that the state deserves preservation.

### Synchronization Collapse

Phase alignment is treated as whole-system coherence.

### Symmetry Collapse

Sameness is treated as organized health.

### Uniformity Collapse

Differentiated contribution is erased in pursuit of identical behavior.

### Connectivity Collapse

A connected network is treated as trustworthy, resilient, or ethical.

### Consensus Collapse

Agreement is treated as truth or justice.

## Entropy Collapse

Coherence is treated as the mathematical negative of thermodynamic or information entropy.

## Efficiency Collapse

Maximum output per unit resource is treated as maximum coherence.

## Performance–Reserve Collapse

Present function is treated as sustainable capacity.

## Balance-as-Fifty–Fifty

Dynamic balance is confused with equal shares among every component.

## Rate–Level Confusion

$$\dot{C} = 0$$

is treated as evidence of high $C$.

## Level–Direction Confusion

A high present $C$ is treated as safe despite rapid decline.

## Margin Concealment

The system's distance from threshold is omitted.

## Debt Concealment

Deferred maintenance is hidden because output remains visible.

## Cost Concealment

Regulatory burden is excluded from the assessment.

## External-Support Concealment

A system appears self-sustaining because unreported external resources continually stabilize it.

## Boundary Manipulation

Losses are excluded by narrowing the system boundary.

## Local–Global Collapse

A subsystem's success is treated as whole-system coherence.

## Robustness Universalism

Robustness to one disturbance is exported to every disturbance.

## Redundancy–Independence Confusion

Several components sharing one failure mode are treated as independent protection.

### Modularity Absolutism

Isolation of modules is treated as universally beneficial.

### Integration Absolutism

Maximum coupling is treated as universally beneficial.

### Resilience Romanticism

The ability to survive harm is used to justify exposing the system to preventable harm.

### Recovery-as-Return

Restoration of the previous state is assumed to be the only legitimate recovery.

### Transformation-as-Failure

A new viable identity is rejected merely because the old form did not return.

### Generation–Goodness Collapse

Any process increasing focal coherence is treated as morally good.

### Loss–Evil Collapse

Every loss process is treated as meaningless, even when release or differentiation is necessary.

### Conflict–Incoherence Collapse

Disagreement is treated as failure even when it produces correction.

### Silence–Coherence Collapse

Absence of visible conflict is treated as trust.

### Measurement-Unit Erasure

Incompatible variables are added without normalization or justification.

### Threshold Arbitrary Authority

An unvalidated threshold is treated as a natural law.

### Time-Window Erasure

A process beneficial over minutes is treated as beneficial over years.

### Path-Dependence Erasure

Returning an input to its prior value is assumed to restore the prior state.

### Model Adequacy Erasure

A regulator's internal model is assumed correct because regulation once succeeded.

### Adaptive-Cost Erasure

The cost of continual adaptation is omitted.

### Learning-Capacity Erasure

The system's present function is praised while its capacity to correct itself is disappearing.

### Memory Worship

Inherited organization is preserved despite stronger evidence.

### Meaning Immunity

A declared purpose is protected from empirical and ethical correction.

### Ethical Externality Erasure

The focal system exports harm while claiming internal coherence.

### Dignity as a Weight

Human standing is treated as one compensable term in an aggregate objective.

### Health-Score Universalism

One physiological rhythm or proprietary index is called whole-person coherence.

### Machine-Coherence Anthropomorphism

Computational stability or synchronized processing is treated as lived integrity.

### Spirit-Field Physicalization

Symbolic interior coherence is converted into an unsupported physical field or force.

## The Coherence Constitution

Before any coherence claim enters Book II, it must answer:

1. What system is being evaluated?
2. Where is its boundary?
3. What lies outside the boundary?
4. What relationships are indispensable?
5. Who declared them indispensable?
6. What identity is to be preserved?
7. What functions are to be preserved?
8. What adaptive capacity is required?
9. What learning capacity is required?
10. What form of participation is required?
11. Is participation merely functional or also ethical?
12. What coherence domain is being discussed?
13. Is the term physical, quantum, phase, network, physiological, psychological, institutional, or architectural?
14. Are several domains being conflated?
15. What is the state space?
16. What are the state variables?
17. What are their units?
18. What dynamics govern them?
19. What boundary conditions apply?

20. What initial conditions apply?
21. What time interval matters?
22. What observation process is used?
23. What uncertainty accompanies observation?
24. Is coherence represented as a vector or scalar?
25. Why is scalar reduction justified?
26. What dimensions enter the vector?
27. What dimensions are excluded?
28. What normalization is used?
29. What thresholds are used?
30. Who selected the thresholds?
31. Are thresholds empirical, legal, ethical, or architectural?
32. What happens at the threshold?
33. Is failure gradual or discontinuous?
34. What is the coherence margin?
35. Is geometric margin being confused with disturbance tolerance?
36. What defines the viable set?
37. Is the viable set time-dependent?
38. Does a viable trajectory exist?
39. What control actions are available?
40. Are those actions causally available in real time?
41. What disturbances are included?
42. What disturbances are excluded?
43. Is robust viability required?
44. What quantifier order governs control and disturbance?
45. What is the coherence variable $C$?
46. Is $C$ dimensionless?
47. What are the units of $G$?
48. What are the units of $L$?
49. Are $G$ and $L$ nonnegative?
50. How are signed process contributions classified?
51. Can a process change sign?
52. What interaction terms exist?
53. Is any process double-counted?
54. Is coherence being treated as conserved?
55. What external generation enters the system?
56. What losses leave the system?
57. Is the system open or closed?
58. Is dynamic balance being claimed?
59. Does dynamic balance mean only $\dot{C} = 0$?
60. What is the actual level of $C$?
61. Is balance being confused with fifty–fifty allocation?
62. Are unequal contributions necessary?
63. Is differentiation being preserved?
64. Is synchronization being measured?
65. What does the synchronization measure mean?
66. Is phase alignment being confused with health or wisdom?
67. Is a graph model used?

68. What do the nodes represent?
69. What do the edges represent?
70. Is the graph directed or undirected?
71. Is it weighted?
72. What does algebraic connectivity reveal?
73. What does it not reveal?
74. Is consensus being sought?
75. Could consensus be false?
76. Could consensus be coerced?
77. Is network centralization creating fragility?
78. Is redundancy present?
79. Are redundant components independent?
80. Is degeneracy present?
81. Is modularity present?
82. Does modularity contain disturbance?
83. Does modularity obstruct learning?
84. Is an order parameter used?
85. How was it derived?
86. What microscopic distinctions does it erase?
87. Is stability being claimed?
88. What reference state or trajectory is used?
89. What norm defines nearness?
90. Is a Lyapunov function available?
91. What does the Lyapunov function measure?
92. Is the stable state desirable?
93. Is asymptotic stability required?
94. Is the reference a fixed point, orbit, attractor, or moving manifold?
95. Is resilience being claimed?
96. Is resilience resistance, basin size, return time, or reorganization?
97. What disturbance magnitude is absorbed?
98. What recovery time is measured?
99. What state is recovered?
100. Should the old state be recovered?
101. What post-disturbance capacity remains?
102. Is robustness being claimed?
103. Robustness to what disturbance class?
104. Robustness of which property?
105. Over what horizon?
106. What new fragility accompanies the robustness?
107. Is random failure distinguished from targeted failure?
108. Is homeostasis being modeled?
109. What variable is regulated?
110. What range is acceptable?
111. Is the reference fixed or adaptive?
112. Is allostasis involved?
113. What anticipated demand changes the reference?
114. What is the cost of regulation?
115. What energy or resource is consumed?
116. What reserve remains?
117. Is reserve measurable?
118. Is coherence debt accumulating?

119. What maintenance has been deferred?
120. Who will bear repayment?
121. Is present output being preserved by consuming future capacity?
122. Is regulatory variety sufficient?
123. What disturbance variety exists?
124. What response variety exists?
125. What residual variety remains?
126. Does the regulator possess a model?
127. How accurate is that model?
128. Is the model changing with the environment?
129. Is the system learning?
130. Can it detect model failure?
131. Can memory preserve obsolete regulation?
132. Can the system unlearn?
133. Is meaning determining what should be preserved?
134. Is the purpose empirically accurate?
135. Is the purpose ethically legitimate?
136. Could a destructive organization become internally coherent?
137. Coherent for whom?
138. Who benefits?
139. Who bears the losses?
140. What externalities are exported?
141. Are future generations represented?
142. Is internal coherence being confused with relational coherence?
143. Is local coherence damaging higher-level coherence?
144. What scales are represented?
145. What mappings connect the scales?
146. Are cross-scale residuals measured?
147. Does one scale dominate every other?
148. Are constraints mutually feasible?
149. What slack remains?
150. Which constraint is nearest to violation?
151. Could the nearest constraint be less dangerous than another dynamic boundary?
152. Is entropy being invoked?
153. Which entropy?
154. Is thermodynamic entropy being confused with information entropy?
155. Is entropy being confused with disorder?
156. Is coherence being treated as negative entropy?
157. Is quantum coherence being invoked?
158. What basis defines it?
159. What operations define the resource theory?
160. Is quantum vocabulary being exported into interior life without evidence?
161. Is stochastic disturbance modeled?
162. What noise process is used?
163. Are jumps or rare events present?
164. What is the first-passage risk?
165. What is the probability of threshold crossing?
166. Are averages concealing brief collapse?
167. What is the minimum coherence?
168. How long is the system below threshold?

169. Is hysteresis present?
170. Can the old state be restored by reversing the input?
171. What history must be repaired?
172. Is psychological coherence being claimed?
173. What interior evidence exists?
174. What behavioral evidence exists?
175. Is conflict being confused with incoherence?
176. Is silence being confused with trust?
177. Is institutional coherence being claimed?
178. Is mission aligned with policy?
179. Is policy aligned with action?
180. Is action accountable?
181. Can the institution learn?
182. Is the mission itself legitimate?
183. Is civilizational coherence being claimed?
184. Are justice, health, education, ecology, and future capacity included?
185. Is economic output being allowed to compensate for noncompensable loss?
186. Is dual-aspect coherence being modeled?
187. What is exterior?
188. What is interior?
189. What is the interface?
190. Are these descriptions being treated as interchangeable?
191. Is the Spirit Field invoked?
192. Is its symbolic status explicit?
193. Is any physical field claim being made?
194. What evidence would be required?
195. What would require the coherence model to change?
196. What would require the purpose to change?
197. What should not be preserved?
198. What loss is necessary for renewal?
199. What remains unknown?
200. What does the coherence claim not establish?

A coherence claim unable to answer these questions may remain useful as:

$$[ARC], [MOD], [SYM], \text{or} [CONJ].$$

Its status must remain visible.

## The Architect's Reflection

For many years, I believed coherence meant holding everything together.
Reality gradually taught another lesson.
Some things should not be held together forever.
A false belief must be released.
A damaged structure may require reconstruction.
An inherited practice may deserve correction.
A relationship may require a stronger boundary.
A purpose may need to change when its consequences become visible.
Coherence is not the refusal to let anything end.

It is the discipline of recognizing what must remain faithful through the ending.

Perhaps the Cathedral has never stood because none of its stones cracked.

Perhaps it stood because builders noticed the cracks.

Because load was redistributed.

Because damaged stones were repaired.

Because one generation returned to work left unfinished by another.

Because the foundation remained more important than the builder's pride.

Book I first named the Cathedral.

Book II has now walked beneath its floor.

It has touched the foundations.

Measured the arches.

Examined the load.

Found the thresholds.

Asked what happens when a stone weakens.

Asked how the Cathedral continues carrying people it will never know.

Perhaps this is why the chapter feels emotional.

The equation:

$$\frac{dC}{dt} = G - L$$

is no longer only an equation.

It is the quiet history of every person who helped us remain ourselves while life continued changing.

Every parent who carried more than an equal share when a child could not.

Every teacher who preserved a possibility a student had not yet recognized.

Every caregiver who supplied coherence while another person's reserves were diminished.

Every ancestor who left one stone.

Every scientist who corrected an inherited error.

Every builder who understood that the Cathedral did not belong to them.

Perhaps coherence is not merely what remains.

Perhaps coherence is the relationship through which what remains becomes worthy of being carried forward.

## Quiet Reflection

Do not begin with a theory.

Think of your own life.

What relationships preserve your identity?

What activities preserve your function?

What allows you to adapt?

What keeps you teachable?

What permits meaningful participation?

Now ask:

What presently generates coherence?

What presently diminishes it?

What appears stable only because reserve is being spent?

What maintenance has been postponed?

What burden are you carrying that once protected you but now consumes you?

What part of your life is nearest its threshold?

What deserves repair?

What deserves a stronger boundary?

What deserves to be released?

Think of the people who have preserved your coherence.

Not by making life perfect.

Not by preventing every loss.

But by helping something essential remain faithful through change.

Remain with one final question:

What must remain—and what must be allowed to change—for my life to remain faithful to itself?

## Constitutional Declaration of Coherence Integrity

We will distinguish coherence from perfection.

We will distinguish coherence from immobility.

We will distinguish coherence from symmetry.

We will distinguish coherence from equilibrium.

We will distinguish coherence from synchronization.

We will distinguish coherence from uniformity.

We will distinguish coherence from connectivity.

We will distinguish coherence from consensus.

We will distinguish coherence from efficiency.

We will distinguish coherence from low entropy.

We will distinguish scientific uses of coherence from architectural uses.

We will preserve the empirical integrity of physical coherence.

We will preserve the basis dependence of quantum coherence.

We will not convert quantum terminology into unsupported claims about consciousness, health, spirituality, or the Spirit Field.

We will declare every system boundary.

We will declare what lies beyond it.

We will declare every indispensable relationship.

We will declare the identity being preserved.

We will declare the function being preserved.

We will declare the adaptive capacity being preserved.

We will declare the learning capacity being preserved.

We will declare the participation being preserved.

We will distinguish internal coherence from relational coherence.
We will distinguish focal success from whole-system success.
We will ask coherent for whom.
We will ask at whose cost.
We will ask across which timescale.
We will preserve Book I's definition.
We will deepen it without pretending it has become one universal empirical law.
We will preserve the Cathedral Equation:

$$\frac{dC}{dt} = G - L.$$

We will declare $C$.

We will declare its units.

We will declare $G$.

We will declare $L$.

We will not treat coherence as a conserved physical quantity.
We will not hide interactions inside unexplained totals.
We will recognize that processes may change sign with context.
We will distinguish level from rate.
We will distinguish high coherence from increasing coherence.
We will distinguish low coherence from declining coherence.

We will not call $\dot{C} = 0$ health.

We will not confuse balance with fifty-fifty equality.
We will recognize organized proportion.
We will preserve differentiated contribution.
We will not require every component to perform the same role.
We will define the viable region.
We will define every threshold.
We will justify every threshold.
We will measure margin.
We will not confuse geometric margin with actual disturbance tolerance.
We will examine viability.
We will examine robust viability.
We will preserve the causal order between disturbance and response.
We will examine stability.
We will not call every stable state desirable.
We will examine attractors.
We will examine basins.
We will examine trajectories.
We will not require every coherent system to remain at one point.
We will permit cycles.

We will permit development.
We will permit transformation.
We will examine resilience.
We will distinguish resistance from recovery.
We will distinguish recovery from return.
We will examine post-disturbance capacity.
We will not romanticize resilience.
We will not use another person's ability to survive harm as permission to expose them to preventable harm.
We will examine robustness.
We will complete every robustness claim with a disturbance class.
We will examine the fragility shadow of every robust architecture.
We will examine random failure.
We will examine targeted failure.
We will examine redundancy.
We will examine whether redundant pathways truly fail independently.
We will examine degeneracy.
We will examine modularity.
We will not maximize integration.
We will not maximize separation.
We will preserve sufficient relationship among differentiated parts.
We will examine homeostasis.
We will not reduce regulation to one fixed setpoint.
We will examine allostasis.
We will examine anticipatory regulation.
We will measure the cost of regulation.
We will measure reserve.
We will detect coherence debt.
We will not praise outward function while future capacity is silently consumed.
We will disclose deferred maintenance.
We will ask who will bear its repayment.
We will examine regulatory variety.
We will not demand that a regulator answer disturbances it cannot distinguish.
We will examine the regulator's model.
We will preserve model uncertainty.
We will revise the model when reality changes.
We will examine phase synchronization.
We will not call maximum synchronization maximum coherence.
We will preserve differentiation.
We will examine network connectivity.
We will not call a connected network truthful.
We will not call consensus justice.

We will preserve dissent capable of correcting error.

We will examine central nodes.

We will examine single points of failure.

We will examine multiscale coherence.

We will not allow local optimization to destroy the larger system.

We will examine cross-scale compatibility.

We will preserve future generations within the system boundary.

We will examine constraints.

We will examine slack.

We will admit when the declared requirements are mutually impossible.

We will revise goals rather than demanding impossible sacrifice from the system.

We will preserve thermodynamic integrity.

We will distinguish entropy flow from entropy production.

We will not identify coherence with negative entropy.

We will not use thermodynamic vocabulary as moral vocabulary.

We will examine uncertainty.

We will model noise where appropriate.

We will examine threshold-crossing risk.

We will not allow averages to erase collapse.

We will measure minimum states.

We will measure time below threshold.

We will examine hysteresis.

We will remember that reversing a disturbance may not reverse its history.

We will preserve repair.

We will preserve maintenance.

We will preserve learning.

We will preserve memory.

We will preserve knowledge.

We will preserve meaning.

We will not preserve inherited error merely because it is old.

We will not discard inherited wisdom merely because it is old.

We will allow reality to distinguish between them.

We will not treat conflict as automatic incoherence.

We will not treat silence as trust.

We will examine whether disagreement strengthens correction or destroys relationship.

We will examine institutional coherence.

We will align mission, policy, action, accountability, and learning.

We will examine the legitimacy of the mission itself.

We will examine civilizational coherence.

We will not allow economic output to compensate silently for lost justice, health, ecology, trust, or future capability.

We will preserve dignity as a constitutional boundary.

We will preserve consent.

We will preserve agency.

We will preserve truth.

We will preserve compassion.

We will distinguish exterior function from interior integrity.

We will distinguish interior experience from exterior measurement.

We will preserve their relationship without collapsing them.

We will preserve the symbolic status of the Spirit Field.

We will not convert gratitude, hope, forgiveness, purpose, wonder, responsibility, or compassion into an unsupported physical force.

We will permit symbolic language to remain meaningful because its status remains honest.

We will recognize that some things should not be preserved.

We will permit falsehood to end.

We will permit harmful structures to change.

We will permit obsolete identities to mature.

We will permit grief to alter the form of continuity.

We will distinguish continuity from repetition.

We will distinguish fidelity from immobility.

We will remember that the Cathedral stands through continual work.

Reality presents disturbance.

Boundary reveals vulnerability.

Regulation answers.

Repair restores.

Adaptation changes.

Learning corrects.

Memory preserves.

Knowledge models.

Meaning chooses what deserves continuation.

Generation builds.

Loss tests.

Reserve prepares.

Debt warns.

Resilience reorganizes.

Coherence keeps the Cathedral standing.

## Crossing the Next Arch

Coherence has now entered mathematics through:

- typed state variables;
- viability regions;

- noncompensable thresholds;
- coherence margins;
- generation and loss rates;
- integral balance;
- signed process contributions;
- vector dynamics;
- repair–loss models;
- viability kernels;
- robust invariance;
- Lyapunov stability;
- trajectories and attractors;
- ecological resilience;
- robustness and fragility;
- homeostasis;
- allostasis;
- regulatory cost;
- reserve;
- coherence debt;
- requisite variety;
- good-regulator models;
- phase synchronization;
- graph Laplacians;
- algebraic connectivity;
- consensus dynamics;
- redundancy;
- degeneracy;
- modularity;
- order parameters;
- multiscale compatibility;
- constraint satisfaction;
- entropy balance;
- stochastic threshold risk;
- hysteresis;
- dual-aspect participation;
- the symbolic Spirit Field;
- the expanded Cathedral Equation.

Yet coherence alone does not answer the final question.

A harmful system may preserve itself coherently.

A false ideology may coordinate action.

A powerful institution may maintain continuity by exporting losses to vulnerable people.

A civilization may possess the capacity to endure while remaining uncertain whether its endurance benefits the future.

Coherence tells us whether organized relationships continue.

It does not by itself tell us:

- which relationships deserve preservation;
- when repair is wiser than release;
- when adaptation becomes surrender;
- when sacrifice becomes exploitation;
- when stability becomes stagnation;
- when transformation becomes responsibility;
- how competing scales should be judged.

The next mathematical question is therefore:

How should knowledge, meaning, uncertainty, power, consequence, and coherence be judged together when no single objective contains the whole of what matters?

Book II must now formalize:

- judgment under uncertainty;
- multi-criteria decision;
- constitutional constraints;
- precaution;
- proportionality;
- stewardship;
- irreversible consequence;
- intergenerational responsibility;
- moral uncertainty;
- collective deliberation;
- the transformation through which coherent understanding becomes wise participation.

The Cathedral is standing.

The final chamber must ask what the Cathedral is for.

## CATHEDRAL THREAD

Before coherence could enter mathematics, reality had to become representable as state.

Before state could preserve identity, relation and differentiation had to become visible.

Before identity could endure, boundaries had to regulate participation.

Before boundaries could matter, vulnerability had to reveal what could be lost.

Before vulnerability could become history, adaptation had to answer disturbance.

Before regulation could become selective, attention had to allocate limited resources.

Before selected information could become experience, awareness had to organize presence.

Before experience could alter the future, learning had to change the learner.

Before learning could endure, memory had to preserve the change.

Before memory could become understanding, knowledge had to organize relationships.

Before knowledge could guide life, meaning had to identify what deserved devotion.

Before meaning could endure through reality's continual change, coherence had to become the Keystone.

The constitutional principle of this chamber is now established:

Coherence is the continued preservation of a declared set of indispensable relationships and capacities within a changing, bounded, vulnerable system.

The typed-coherence principle follows:

Every coherence claim must declare its domain, variable, boundary, scale, and time horizon.

The multidimensional principle follows:

Identity, function, adaptation, learning, and meaningful participation should remain distinguishable until evidence justifies their compression.

The Cathedral Equation follows:

$$\frac{dC}{dt} = G - L$$

The rate principle follows:

Coherence increases when declared generation exceeds declared loss and decreases when loss exceeds generation.

The level-rate principle follows:

A stationary coherence variable may remain high, low, sustainable, or near collapse; rate alone does not reveal condition.

The proportionality principle follows:

Dynamic balance is not universal fifty-fifty equality; coherent systems preserve differentiated contributions in sufficient proportion.

The viability principle follows:

Coherence may consist not in occupying one ideal state, but in retaining a viable path through changing conditions.

The stability principle follows:

Stability describes persistence near a reference condition without establishing that the reference condition deserves preservation.

The resilience principle follows:

Resilience includes disturbance absorption, basin structure, recovery, reorganization, and post-disturbance capacity rather than return speed alone.

The robustness principle follows:

Every robustness claim must name the property preserved and the perturbations against which it is preserved.

The regulatory principle follows:

A regulator must possess enough differentiated response and sufficiently faithful modeling to answer the variety reality presents.

The reserve principle follows:

Present function may conceal declining reserve, rising cost, and accumulating coherence debt.

The synchronization principle follows:

Temporal alignment can support coherence without becoming identical to coherence or justifying the erasure of differentiation.

The network principle follows:

Connectivity and consensus organize relation, but neither guarantees truth, resilience, dignity, or justice.

The multiscale principle follows:

A subsystem cannot claim complete coherence by preserving itself through the progressive destruction of the systems upon which it depends.

The entropy principle follows:

Thermodynamic entropy remains a physical quantity governed by its own laws; SDAE coherence is not its universal negative.

The stochastic principle follows:

Coherence includes not only average condition, but the probability, timing, and consequence of crossing critical boundaries.

The transformation principle follows:

Faithful continuity may require a new viable form rather than restoration of the previous state.

The ethical principle follows:

Internal coordination becomes coherent participation only when externalities, dignity, consent, justice, and future consequence remain within the boundary of responsibility.

The dual-aspect principle follows:

Exterior function, interior integrity, and the boundary joining them may be modeled as related dimensions without becoming interchangeable descriptions.

The Spirit Field principle follows:

Interior coherence may be symbolically organized through understanding, gratitude, hope, responsibility, forgiveness, wonder, purpose, and compassion without being converted into an unearned physical mechanism.

The Book I principle now gathers its full life:

Book I taught the reader to recognize the Keystone. Book II has now revealed the load it carries.

The wisdom principle approaches:

Coherence preserves what continues; wisdom must judge what deserves continuation and how that continuation should serve the whole.

# THE CATHEDRAL EQUATION



*Coherence is organized persistence through continual change.*

**STATIONARITY != STABILITY | COHERENCE != MORAL WORTH**

FS-G19.01 — EQ-22.01 · EQ-22.02

### COHERENCE PASSPORT AND TYPED FUNCTIONAL

Declare system, function, boundary, scale, horizon, context, units, uncertainty, and the Cathedral relation.

FS-G19.02 — EQ-22.01 · EQ-22.04

### DIMENSIONAL AND INTEGRAL BALANCE

Require compatible quantities and distinguish instantaneous rate from accumulated history.

FS-G19.03 — EQ-22.03

### GENERATION-LOSS DECOMPOSITION

Separate generating, maintaining, repairing, adapting, and loss-producing processes.

FS-G19.04 — EQ-22.05 · EQ-22.12 · EQ-22.14

### VECTOR PROFILE AND VIABILITY

Protect critical dimensions, viability margins, and a multidimensional quality profile from scalar erasure.

FS-G19.05 — EQ-22.08 · EQ-22.11

### RESERVE, DEFICIT, HIDDEN BURDEN

Track reserve drawdown, cumulative deficit, exported loss, delayed loss, and masked decline.

FS-G19.06 — EQ-22.06

### DYNAMIC BALANCE AND STABILITY

Distinguish active stationarity, equilibrium, local stability, and finite-perturbation safety.

FS-G19.07 — EQ-22.03 · EQ-22.09

### MAINTENANCE, REPAIR, ADAPTATION

Make maintenance, damage, repair, learning, and adaptive response visible beneath persistence.

FS-G19.08 — EQ-22.09

### ROBUSTNESS, RECOVERY, RESILIENCE

Declare disturbance family, depth, duration, recovery time, hysteresis, and sustained return to viability.

FS-G19.09 — EQ-22.06 · EQ-22.09

### THRESHOLD, TIPPING, TRANSFORMATION

Separate bounded strain, threshold crossing, collapse, reorganization, transformation, and renewal.

FS-G19.10 — EQ-22.10 · EQ-22.11

### MULTISCALE COUPLING AND EXTERNALITY

Audit local support, global burden, cross-scale trade-off, imported reserve, and future consequence.

FS-G19.11 — EQ-22.07

### ENTROPY, INFORMATION, MORAL BOUNDARIES

Keep architectural coherence distinct from thermodynamic entropy, information measures, and moral goodness.

FS-G19.12 — EQ-22.13 · EQ-22.14

### PREDICTION, QUALITY, REVISION

Test predictive gain against serious baselines; preserve complete audit, quality profile, wisdom, and revision.

**DIGNITY | UNCERTAINTY | SOURCE TRANSPARENCY | FAILURE | REVISION**

FS identifies figure position; EQ identifies canonical equation. Reality remains the measure.

CHAPTER TWENTY-THREE

# WISDOM: JUDGMENT, STEWARDSHIP, AND THE MATHEMATICS OF RESPONSIBLE PARTICIPATION

*The Fourteenth Mathematical Chamber*

**Knowledge explains reality.**
**Wisdom learns how to participate faithfully within it.**

**Coherence keeps the Cathedral standing.**
**Wisdom asks what the Cathedral must shelter, what it must repair, and what it must pass forward.**

## Guiding Question

How can mathematics assist judgment when knowledge remains incomplete, values remain plural, consequences remain uncertain, harms may be irreversible, responsibility is distributed, and future generations cannot yet speak—and how can decision theory, risk analysis, robustness, precaution, proportionality, accountability, stewardship, compassion, and intergenerational viability support wisdom without reducing dignity, conscience, moral worth, or responsible participation to optimization?

Chapter Ten established state.

Chapter Eleven established relation.

Chapter Twelve established differentiation.

Chapter Thirteen established boundary.

Chapter Fourteen established vulnerability.

Chapter Fifteen established adaptation.

Chapter Sixteen established attention.

Chapter Seventeen established awareness.

Chapter Eighteen established learning.

Chapter Nineteen established memory.

Chapter Twenty established knowledge.

Chapter Twenty-One established meaning.

Chapter Twenty-Two established coherence.

The Mathematical Cathedral has now reached its final chamber.

A system can occupy a state.

Its parts can relate.

Its identity can differentiate.

Its boundary can regulate exchange.

Its vulnerability can reveal what may be lost.

Its adaptive architecture can answer disturbance.

Its attention can select.

Its awareness can experience.

Its learning can reorganize.

Its memory can preserve.

Its knowledge can understand.

Its meaning can orient.

Its coherence can sustain indispensable relationships through change.

Yet none of these achievements, alone, determines what should now be done.

A system may understand reality and exploit it.

It may preserve itself by transferring loss to others.

It may optimize one objective while destroying everything that the objective omitted.

It may act confidently upon incomplete evidence.

It may refuse to act until all uncertainty disappears.

It may protect the present by narrowing the future.

It may preserve an inherited institution whose purpose has become harmful.

It may produce a technically successful outcome through an ethically incoherent process.

The final question is therefore neither merely scientific nor merely mathematical.

It is constitutional.

**How shall organized understanding become responsible participation?**

## The Return of Book I's Final Chamber

Book I named Wisdom **the highest expression of coherent understanding**.

It declared:

**Wisdom is coherent understanding faithfully lived.**

And it preserved the constitutional sequence:

$$\text{Knowledge explains. Meaning orients. Wisdom participates.}$$

Book I did not identify wisdom with intelligence, information, age, authority, or certainty. It described wisdom as the mature conversion of understanding into life and placed compassion, stewardship, and civilization beyond coherence's Keystone:

$$\text{Reality} \to \text{Observation} \to \text{Science} \to \text{Mathematics} \to \text{Organization} \to \text{Coherence} \to \text{Wisdom} \to \text{Compassion} \to \text{Stewardship} \to \text{Civilization.}$$

Nothing was skipped.

Nothing was assumed.

Everything was inherited.

Everything was earned.

Book I then gave its final constitutional principle:

**Coherence reaches its highest expression when organized understanding continually strengthens life, dignity, responsibility, and future inheritance.**

It recognized stewardship as organized responsibility toward what each generation inherits—Earth, knowledge, mathematics, language, culture, science, and children—and civilization as organized coherence extended through teaching across generations.

Book II now asks what mathematics can contribute to that final movement.

Mathematics can help reveal:

- alternatives;
- uncertainty;
- probability;
- model disagreement;
- expected consequence;
- worst-case loss;
- regret;
- tail risk;
- irreversibility;
- option preservation;
- value of information;
- distribution of burden;
- causal contribution;
- resource constraints;
- intergenerational trajectories.

Mathematics cannot, by numerical assignment alone, determine:

- whose dignity matters;
- which harms are impermissible;
- whether consent is genuine;
- what a life is worth;
- what humanity owes the future;
- whether compassion should enter the boundary;
- what civilization is for.

The boundary of the final chamber is therefore:

$$\text{Mathematical Decision} \neq \text{Wise Judgment}$$

Mathematics can discipline judgment.

It cannot replace the responsibility to judge.

## Terminological Discipline

| Term | Meaning in Book II |
|---|---|
| **Wisdom** | Coherent understanding faithfully lived through reality-accountable, compassionate, proportionate, and responsible participation |
| **Practical wisdom** | Judgment concerning action under variable, contextual, ethically consequential conditions |
| **Theoretical understanding** | Knowledge directed primarily toward explanation, description, or truth |
| **Judgment** | Context-sensitive integration of evidence, uncertainty, values, constraints, relationships, and consequences |
| **Decision** | Selection among available alternatives |
| **Action** | A realized intervention or participation capable of changing the world |
| **Policy** | A retained rule or strategy governing repeated decisions |
| **Deliberation** | Examination of alternatives, reasons, uncertainties, values, and consequences before commitment |
| **Decision architecture** | The declared relationship among alternatives, information, models, values, constraints, and consequences |
| **Decision quality** | Quality of the reasoning and process available before the uncertain outcome is known |
| **Outcome quality** | Quality of what actually occurred after uncertainty was resolved |
| **State of nature** | A condition not fully controlled by the decision-maker that influences consequences |
| **Risk** | Uncertainty represented through a probability model over relevant consequences |
| **Ambiguity** | Uncertainty concerning which probability model should be used |
| **Deep uncertainty** | Condition in which parties cannot confidently agree upon models, probabilities, values, or important future conditions |
| **Epistemic uncertainty** | Uncertainty arising from incomplete knowledge, evidence, or model adequacy |
| **Aleatoric uncertainty** | Variability represented as inherent within the declared model |
| **Moral uncertainty** | Uncertainty concerning which ethical framework, principle, or priority should govern action |
| **Loss** | A numerical penalty representing declared undesirable consequence |
| **Harm** | Injury, deprivation, violation, or burden; not necessarily reducible to one loss function |
| **Benefit** | Improvement or protected interest under a declared evaluative framework |
| **Expected loss** | Probability-weighted average loss under a declared model |
| **Worst-case loss** | Largest modeled loss over a declared uncertainty set |
| **Regret** | Difference between the loss of a chosen action and the best action that could have been chosen had the state been known |
| **Tail risk** | Risk concentrated in severe, comparatively low-probability outcomes |

| Term | Meaning in Book II |
|---|---|
| **Robustness** | Adequate performance across a declared range of models or futures |
| **Satisficing** | Selecting an alternative meeting declared aspiration and constraint levels rather than proving global optimality |
| **Value of information** | Expected improvement in decision quality made possible by obtaining information before commitment |
| **Precaution** | Proportionate preventive action when credible threats of serious or irreversible harm remain despite incomplete certainty |
| **Proportionality** | Requirement that an intervention pursue a legitimate purpose, be suitable, avoid unnecessary restriction, and not impose excessive burden |
| **Irreversibility** | Loss of practical ability to restore an earlier state or recover important future options |
| **Option value** | Value of preserving future choices while uncertainty remains unresolved |
| **Dignity** | Inherent standing that must not be reduced to productivity, convenience, preference, probability, or aggregate utility |
| **Consent** | Voluntary and sufficiently informed authorization by a person possessing relevant decision-making capacity |
| **Autonomy** | Capacity for self-endorsed choice within real rather than merely formal alternatives |
| **Stakeholder** | Person, community, system, or future population materially affected by a decision |
| **Responsibility** | Relationship among causal participation, knowledge, control, role, benefit, and duty to repair |
| **Accountability** | Obligation to explain decisions, preserve records, permit review, correct error, and answer for consequence |
| **Foreseeability** | Degree to which an outcome could reasonably have been anticipated from available evidence |
| **Causal contribution** | Degree to which an agent or action participated in producing an outcome under a declared causal model |
| **Stewardship** | Temporary custodianship of inherited relationships, resources, knowledge, institutions, and future possibilities |
| **Trusteeship** | Governance of something not owned solely for the trustee's immediate benefit |
| **Inheritance** | What one generation, institution, or person receives before contributing anything of its own |
| **Commons** | Resource or inheritance whose use by several participants requires organized governance |
| **Intergenerational responsibility** | Present obligation toward persons and systems that will inherit the consequences of present action |
| **Sustainability** | Preservation of viable future capability rather than present output alone |
| **Compassion** | Recognition through which another being's vulnerability enters one's own field of responsibility |
| **Justice** | Institutional and relational ordering concerning rights, burdens, benefits, standing, and fair participation |
| **Civilization** | Organized inheritance through which knowledge, dignity, justice, compassion, and capability are transmitted across generations |

| Term | Meaning in Book II |
|---|---|
| **Wise participation** | Action that remains faithful to evidence, uncertainty, dignity, relationship, consequence, and future inheritance |

These distinctions must remain visible.

$$\text{Knowledge} \neq \text{Wisdom} \quad \text{Intelligence} \neq \text{Judgment}$$
$$\text{Good Outcome} \neq \text{Good Decision}$$
$$\text{Expected Benefit} \neq \text{Moral Permission}$$
$$\text{Low Probability} \neq \text{Negligible Responsibility}$$
$$\text{Uncertainty} \neq \text{Permission to Ignore} \quad \text{Precaution} \neq \text{Paralysis}$$
$$\text{Robustness} \neq \text{Wisdom}$$
$$\text{Causal Contribution} \neq \text{Complete Moral Blame}$$
$$\text{Stewardship} \neq \text{Ownership}$$

## Scientific, Human, and Mathematical Foundations

This chamber stands upon the inheritance of:

- philosophy;
- medicine;
- psychology;
- neuroscience;
- history;
- anthropology;
- sociology;
- ethics;
- probability;
- statistics;
- decision theory;
- game theory;
- control theory;
- operations research;
- risk analysis;
- causal inference;
- social-choice theory;
- governance;
- sustainability science.

Aristotle distinguished contemplation of invariant matters from deliberation about variable matters that may be otherwise. In his account of practical action, good choice joins truthful reasoning concerning an end with rightly ordered desire and character. The ancient concept of *phronesis* therefore concerns more than calculation: it concerns practical judgment in circumstances where action, character, and ends participate together.

The Berlin Wisdom Paradigm later approached wisdom empirically as expertise concerning the fundamental pragmatics of life. Its criteria included rich factual and procedural knowledge, lifespan contextualism, recognition of plural values and priorities, and management of uncertainty. Its authors explicitly warned that no single empirical method exhausts the complete meaning of wisdom.

Research on reasoning about social conflict has likewise operationalized aspects of wise reasoning through perspective-taking, recognition of uncertainty and change, attention to competing interests, and search for compromise. Such findings do not produce one universal wisdom scale, but they support the distinction between wisdom and raw intelligence or confidence.

Herbert Simon challenged models that assume decision-makers possess complete information, unlimited computational ability, and perfectly organized preferences. He developed alternatives compatible with limited knowledge and finite capacity, including aspiration levels and selection of satisfactory rather than globally proven optimal actions.

Decision theory can formalize alternatives, probabilities, values, losses, and information. Robust-decision methods address cases where long-term predictions and agreed probability models are unavailable by testing strategies across many plausible futures and emphasizing adaptive actions that perform acceptably across them.

Risk analysis also distinguishes average loss from tail loss. Rockafellar and Uryasev developed Conditional Value-at-Risk as a measure capable of representing the severity of losses beyond a chosen quantile and provided an optimization representation useful under general loss distributions.

International ethical frameworks provide another inheritance. UNESCO's Universal Declaration on Bioethics and Human Rights places human dignity, human rights, individual welfare, benefit and harm, autonomy, consent, justice, solidarity, and future generations within the governance of science and technology. It explicitly states that the interests and welfare of the individual should take priority over the sole interest of science or society.

The Rio Declaration's precautionary formulation states that threats of serious or irreversible environmental damage should not be ignored merely because full scientific certainty is absent, while also requiring proportionate and cost-conscious measures. The legal status and exact scope of precaution remain debated, which is itself a reason to state every precautionary model carefully.

UNESCO's Declaration on the Responsibilities of Present Generations toward Future Generations treats Earth, ecosystems, resources, cultural heritage, freedom of choice, dignity, peace, and education as inheritances that present generations must not irreversibly compromise.

These inheritances do not yield one equation of wisdom.

They reveal the formal burden wisdom must carry.

## Architectural Recognition

Only after science, mathematics, history, ethics, and lived experience have spoken does the Architecture introduce its final recognition:

**Wisdom is coherent judgment under uncertainty through which knowledge and meaning become compassionate, proportionate, accountable, and future-aware participation within reality.**

A general wisdom architecture may be represented as:

$$W = (K, A, \Theta, M, V, H, R, F, E) [ARC/MOD]$$

where:

- $K$ is organized knowledge;
- $A$ is the set of genuinely available actions;
- $\Theta$ is the uncertainty space;
- $M$ is the collection of plausible models;
- $V$ is the value and meaning architecture;
- $H$ is the harm and dignity architecture;
- $R$ is the relational and responsibility architecture;
- $F$ is the future and inheritance architecture;
- $E$ is the ethical and evidentiary boundary.

Wisdom is not located in any one component.

It appears through their organization.

## Judgment Begins Before Optimization

Let:

$$\Theta$$

be a set of possible states of reality.

Let:

$$A$$

be the set of actions.

Let:

$$D$$

be available evidence.

Let:

$$M$$

be the model connecting actions and states to outcomes.

Let:

$$Y = F(a, \theta, M)$$

be the resulting outcome.

Let:

$$l(a,\theta) = \begin{bmatrix} l_1(a,\theta) \\ \vdots \\ l_m(a,\theta) \end{bmatrix}$$

be a vector of declared losses.

Let:

$$h_j(a,\theta) \leq 0$$

represent noncompensable constraints.

The admissible action set is:

$$A_{adm} = \{a \in A : h_j(a, \theta) \leq 0 \text{ under the declared standard}\}.$$

Only after this architecture is declared may optimization begin.

## Equation Passport 23.1

*Decision Architecture Under Uncertainty*

$$Y = F(a, \theta, M) \, l(a, \theta) = (l_1, \ldots, l_m)^T \, A_{adm} = \{a : h_j(a, \theta) \leq 0\}$$

**Status:** $[¿/MOD]$

**Action:** $a$

**Uncertain state:** $\theta$

**Model:** $M$

**Outcome:** $Y$

**Loss vector:** $l$

**Constitutional constraints:** $h_j$

**Required declarations:**

- who chooses;
- who is affected;
- what alternatives are real;
- what uncertainty is represented;
- what uncertainty is omitted;
- which values enter the losses;
- which values remain constraints;
- what time horizon applies.

**What it establishes:** A typed decision problem.

**What it does not establish:** Which ethical architecture should govern it.

**Architectural significance:** Wisdom begins by refusing to optimize an undeclared problem.

# A Good Outcome Can Follow a Poor Decision

Suppose the decision-maker chooses:

$$a_t$$

using evidence:

$$D_t.$$

Before the uncertain state is revealed, decision quality may be assessed through:

$$Q_{dec} = -E\left[L\left(a_t, \Theta\right) \mid D_t\right].$$

After the state:

$$\theta^{¿}$$

occurs, outcome quality is:

$$Q_{out} = -L\left(a_t, \theta^{¿}\right).$$

A reckless action may produce a good outcome through luck.

A careful action may produce a poor outcome because a low-probability event occurred.

Therefore:

$$Q_{dec} \neq Q_{out}.$$

## Equation Passport 23.2

***Decision Quality and Outcome Quality***

$$Q_{dec} = -E\left[L\left(a, \Theta\right) \mid D\right] Q_{out} = -L\left(a, \theta^{¿}\right)$$

**Status:** $\left[¿ / MOD\right]$

**Ex ante quality:** $Q_{dec}$

**Realized quality:** $Q_{out}$

**What a decision review should preserve:**

- evidence available at the time;
- uncertainty known at the time;
- alternatives available at the time;

- values and constraints used;
- outcome eventually observed.

**Failure condition:** Hindsight replaces fair evaluation of the earlier decision.

**Architectural significance:** Wisdom learns from outcomes without pretending that luck and judgment are identical.

## Bounded Rationality

No person or institution can:

- enumerate every possible future;
- calculate every consequence;
- know every probability;
- compare every value;
- investigate indefinitely.

The considered action set is therefore often:

$$A_t^{considered} \subsetneq A .$$

Let:

$$q_j(a)$$

represent achievement in dimension $j$.

Let:

$$\tau_j$$

be an aspiration threshold.

Let:

$$h_k(a) \leq 0$$

be a constitutional constraint.

The satisfactory set is:

$$S_t = \{a, \in A_t^{considered} : q_j(a) \geq \tau_j \forall j \, h_k(a) \leq 0 \forall k\}.$$

A satisficing procedure may select:

$$a_t^{\circ} = FirstAcceptable(S_t)$$

or choose the least costly member of $S_t$.

## Equation Passport 23.3

*Satisficing Judgment*

$$S_t = \{a, : q_j(a) \geq \tau_j \;\; h_k(a) \leq 0\}$$

**Status:** $[MOD]$

**Considered alternatives:** $A_t^{considered}$

**Aspiration levels:** $\tau_j$

**Constraints:** $h_k$

**Satisfactory alternatives:** $S_t$

**Possible strengths:**

- finite search;
- timely action;
- preservation of constitutional minima;
- compatibility with limited cognitive resources.

**Possible failures:**

- aspiration level too low;
- alternatives excluded prematurely;
- first satisfactory option hides a much safer one;
- inherited constraints preserve injustice;
- search order creates bias.

**Architectural significance:** Wisdom need not prove that one action is globally best; it must justify why the action is sufficiently faithful under real limits.

# Wisdom Does Not Worship Optimization

Optimization asks:

Which admissible action minimizes or maximizes the declared objective?

Wisdom asks additional questions.

Was the objective correctly chosen?

Were the affected people represented?

Was consent genuine?

Were long-term consequences included?

Were dignity constraints preserved?

Did the model exclude an important future?

Was the action reversible?

Who bears the uncertainty?

Who receives the benefit?

Who remains able to correct the decision?

An optimizer can solve the wrong problem perfectly.

## Expected-Loss Judgment

Suppose a probability distribution:

$$P(\theta \mid D)$$

is available.

A Bayesian decision rule is:

$$a_B^{*} = \arg\min_{a \in A_{adm}} E[L(a, \Theta) \mid D].$$

This rule is coherent inside the declared probability and loss architecture.

It becomes fragile when:

- the model is wrong;
- probabilities are unstable;
- rare harm is understated;
- the loss function omits dignity;
- affected people reject the valuation;
- future generations are excluded.

## Worst-Case Judgment

When probabilities are unavailable or distrusted, one may use:

$$a_{MM}^{\text{¿}} = arg \min_{a \in A_{adm}} {}^{\text{¿}} \theta \in \Theta \, L(a, \theta).$$

This minimizes worst-case modeled loss.

It can protect against severe outcomes.

It may also become excessively conservative when one implausible scenario dominates every choice.

## Regret

Define statewise best achievable loss:

$$L^{\text{¿}}(\theta) = \min_{b \in A_{adm}} L(b, \theta).$$

Regret is:

$$Reg(a, \theta) = L(a, \theta) - L^{\text{¿}}(\theta).$$

The minimax-regret action is:

$$a_{MR}^{\text{¿}} = arg \min_{a \in A_{adm}} {}^{\text{¿}} \theta \in \Theta \, Reg(a, \theta).$$

Savage's statistical-decision work helped formalize regret-based alternatives to ordinary expected-loss and minimax criteria.

## Equation Passport 23.4

*Expected Loss, Worst-Case Loss, and Regret*

$$a_B^{¿} = \underset{a}{arg\,min}\, E[L(a,\Theta)] a_{MM}^{¿} = \underset{a}{arg\,min}^{¿}\, \theta\, L(a,\theta)$$

$$a_{MR}^{¿} = \underset{a}{arg\,min}^{¿}\, \theta \left[ L(a,\theta) - \underset{b}{min}\, L(b,\theta) \right]$$

**Status:** $[¿/MOD]$

**Bayesian rule:** Requires a probability model

**Minimax rule:** Protects against modeled worst loss

**Minimax regret:** Protects against being far from statewise best

**No universal winner:** Each expresses a different judgment toward uncertainty

**Failure condition:** A decision criterion is treated as value-neutral.

**Architectural significance:** Decision rules do not eliminate judgment; they formalize different attitudes toward what uncertainty should be allowed to do.

# Deep Uncertainty

Sometimes disagreement extends beyond probabilities.

Participants may disagree about:

- which model is correct;
- which variables matter;
- which futures are plausible;
- which values deserve priority;
- which thresholds are legitimate.

Let:

$$P = \{P_1, \ldots, P_K\}$$

be plausible probability models.

Let:

$$M = \{M_1, \ldots, M_R\}$$

be plausible structural models.

A distributionally robust action is:

$$a_{rob}^{¿} = arg \min_{a \in A_{adm}} {}^{¿} P \in P E_P [L(a, \Theta)].$$

A scenario-robust alternative may maximize the proportion of futures in which declared requirements are met:

$$a_{sat}^{¿} = arg \max_{a} \frac{1}{K} \sum_{k=1}^{K} 1\{q_j(a, \theta_k) \geq \tau_j \forall j\}.$$

Robust-decision methods do not prove one future or one strategy correct. They search for actions whose justification survives across a broad ensemble of plausible futures and often favor adaptive strategies that can change as evidence arrives.

## Equation Passport 23.5

***Robust Judgment Across Models***

$$a_{rob}^{¿} = arg \min_{a} {}^{¿} P \in P E_P [L(a, \Theta)]$$

**Status:** $[MOD]$

**Plausible models:** $P$

**Action:** $a$

**Loss:** $L$

**Robustness target:** Adequate performance across model disagreement

**Required declarations:**

- how the model set was constructed;
- whether dissenting models were excluded;
- what counts as adequate;
- which stakeholders define failure;
- whether adaptation remains possible.

**What robustness does not establish:**

- moral goodness;

- truth of any model;
- immunity from unknown unknowns;
- universal optimality.

**Architectural significance:** Wisdom sometimes seeks an action whose reasons remain standing even when the forecast changes.

## Acting and Waiting

Uncertainty does not always support immediate action.

Nor does it always support delay.

The decision depends upon:

- expected value of additional information;
- cost of obtaining it;
- cost of delay;
- reversibility;
- urgency;
- risk of irreversible harm.

Let baseline risk be:

$$R_0 = \min_a E\left[L(a, \Theta)\right].$$

If the state could be known perfectly before action:

$$R_{PI} = E_\Theta\left[\min_a L(a, \Theta)\right].$$

The expected value of perfect information is:

$$EVPI \text{ ¿ } R_0 - R_{PI} \geq 0.$$

For imperfect information $Y$:

$$R_Y = E_Y\left[\min_a E\left[L(a, \Theta) \mid Y\right]\right].$$

Then:

$$EVSI \text{ ¿ } R_0 - R_Y.$$

Howard's information-value theory formalized how information can be valued through the improvement it enables in a decision rather than through information quantity alone.

### Equation Passport 23.6

*Value of Information*

$$EVPI ¿ \min_a E[L(a,\Theta)] - E_\Theta\left[\min_a L(a,\Theta)\right] EVSI ¿ R_0 - R_Y$$

**Status:** $[¿/MOD]$

**Baseline risk:** $R_0$

**Risk with perfect information:** $R_{PI}$

**Risk with sample information:** $R_Y$

**Information cost:** $C_I$

**Nominal acquisition condition:**

$$EVSI ¿ C_I.$$

**Additional wisdom conditions:**

- delay does not create greater harm;
- information gathering respects consent;
- information can alter the action;
- the model can use the information correctly.

**Architectural significance:** Wisdom knows that more information is valuable only when it can arrive in time, be obtained justly, and change what should be done.

## Information Can Have Negative Practical Value

Information may:

- arrive too late;
- distract attention;
- violate privacy;
- create false confidence;
- reveal identities unnecessarily;
- be misunderstood;
- trigger harmful action.

A test whose result cannot change management may add burden without improving judgment.

The value of information is decision-relative.

$$\text{Information Value} \neq \text{Information Quantity}.$$

## Average Risk Is Not Enough

Suppose loss is random variable:

$$Z = L(a, \Theta).$$

Its expectation is:

$$E[Z].$$

Two actions can possess equal expected loss while differing greatly in catastrophic possibility.

Value-at-Risk at level $\alpha$ is:

$$VaR_\alpha(Z) = inf\{z : P(Z \le z) \ge \alpha\}.$$

A Conditional Value-at-Risk representation is:

$$CVaR_\alpha(Z) = \min_{\eta \in R} ¿$$

CVaR represents the severity of the upper loss tail rather than merely locating its quantile.

### Equation Passport 23.7

*Tail-Risk Judgment*

$$VaR_\alpha(Z) = inf\{z : P(Z \le z) \ge \alpha\} CVaR_\alpha(Z) = \min_{\eta} ¿$$

**Status:** $[¿ / MOD]$

**Loss:** $Z$

**Confidence level:** $\alpha$

**VaR:** Quantile threshold

**CVaR:** Average severity in the modeled tail

**What tail analysis adds:** Visibility of severe outcomes hidden by averages

**What it does not determine:**

- whether the loss is morally permissible;
- whose losses enter $Z$;
- whether probabilities are reliable;
- whether catastrophe is compensable.

**Architectural significance:** Wisdom does not permit an attractive average to conceal who may be destroyed in the tail.

## Precaution

Let:

$$H(a,\theta)$$

measure a serious-harm condition, where:

$$H(a,\theta)>0$$

indicates violation.

A chance constraint is:

$$P(H(a,\Theta)\leq 0)\geq 1-\varepsilon .$$

A robust constraint is:

$$\underset{\theta \in \Theta}{¿} H(a,\theta)\leq 0.$$

A tail-risk constraint may be:

$$CVaR_{\alpha} ¿$$

The Rio precautionary formulation applies where serious or irreversible environmental damage is credibly threatened despite incomplete scientific certainty. It does not define one universal formula, and its scope remains contested.

## Equation Passport 23.8

*Precautionary Constraint*

$$P(H(a,\Theta)\leq 0)\geq 1-\varepsilon$$

or:

$$\underset{\theta \in \Theta}{¿} H(a,\theta)\leq 0$$

**Status:** $[MOD/ARC]$

**Harm function:** $H$

**Tolerance:** $\varepsilon$

**Tail threshold:** $\kappa$

**Precaution becomes stronger when:**

- harm is severe;
- harm is irreversible;
- affected people are vulnerable;
- uncertainty is structural;
- later repair is impossible;
- safer alternatives exist.

**Precaution becomes weaker when:**

- harm is minor and reversible;
- delay creates greater harm;
- the feared pathway lacks plausible evidence;
- prevention is itself disproportionate.

**Architectural significance:** Uncertainty should not always reduce responsibility; under irreversible danger, uncertainty may enlarge the duty to preserve options and prevent harm.

# Precaution Is Not Fear

Precaution should not mean:

- prohibit everything unfamiliar;
- demand zero risk;
- ignore benefits;
- treat speculation as evidence;

- protect existing power from innovation.

A precautionary judgment must remain:

- evidence-aware;
- proportionate;
- revisable;
- transparent;
- attentive to the harms of action and inaction.

The Status Quo Also Carries Risk.

# Irreversibility

Let the system evolve as:

$$x_{t+1} = F(x_t, a_t, \theta_t).$$

Let:

$$\Omega_t$$

be the set of viable future options available at time $t$.

After action $a$:

$$\Omega_{t+1}(a)$$

is the remaining option set.

Define an illustrative option-retention ratio:

$$O(a) = \frac{\mu(\Omega_{t+1}(a))}{\mu(\Omega_t)}$$

when a meaningful measure $\mu$ exists.

An action is strongly option-destroying when:

$$O(a) \ll 1.$$

Let return cost be:

$$C_{return}(a) = \inf_{\pi} \{\text{cost of policy } \pi \text{ restoring the prior viable region}\}.$$

A process is practically irreversible when:

$$C_{return}(a)$$

is infinite, infeasible, or unacceptably large.

### Equation Passport 23.9

*Irreversibility and Option Preservation*

$$O(a) = \frac{\mu(\Omega_{t+1}(a))}{\mu(\Omega_t)} C_{return}(a) = \inf_{\pi} \{\text{return cost}\}$$

**Status:** $[MOD/ARC]$

**Present option set:** $\Omega_t$

**Future option set:** $\Omega_{t+1}(a)$

**Option retention:** $O(a)$

**Return cost:** $C_{return}$

**Possible irreversible losses:**

- life;
- species;
- ecosystem function;
- cultural heritage;
- privacy after disclosure;
- trust after betrayal;
- opportunity after critical delay.

**Failure condition:** Numerical option count treats trivial and indispensable options as equivalent.

**Architectural significance:** Wisdom preserves future freedom when present certainty is insufficient to justify closing every door.

## Reversibility Has Value

When two actions produce similar present benefit, the more reversible action may be wiser because it preserves:

- correction;
- learning;
- consent withdrawal;
- alternative futures;
- institutional trust.

This does not mean every decision should remain temporary.

Some commitments require durability.

The question is:

**Has irreversibility been earned by evidence, legitimacy, and necessity?**

## Proportionality

A proportionate intervention should satisfy several distinguishable tests.

Let:

$$P(a) = \begin{bmatrix} L_g(a) \\ E_f(a) \\ N_c(a) \\ B_l(a) \\ R_h(a) \end{bmatrix}$$

where:

- $L_g$ is legitimacy of aim;
- $E_f$ is likely effectiveness;
- $N_c$ is necessity relative to less restrictive alternatives;
- $B_l$ is balance between benefit and burden;
- $R_h$ is compliance with rights and dignity.

A proportionate action requires:

$$L_g(a) = 1, E_f(a) \geq E_{min},$$

no materially less restrictive alternative $b$ satisfying:

$$E_f(b) \approx E_f(a)$$

with:

$$H(b) < H(a),$$

and all constitutional rights constraints preserved.

UNESCO's AI ethics recommendation uses proportionality, contextual assessment, stakeholder engagement, due diligence, and impact assessment to govern conflicts among desirable values and insists that limitations on rights be lawful, necessary, reasonable, and proportionate.

## Equation Passport 23.10

*Proportionality Architecture*

$$P(a) = (L_g, E_f, N_c, B_l, R_h)^T$$

**Status:** $[ARC/MOD]$

**Legitimate aim:** $L_g$

**Effectiveness:** $E_f$

**Necessity:** $N_c$

**Balance:** $B_l$

**Rights compatibility:** $R_h$

**What it prevents:** Treating "it works" as sufficient justification

**Failure conditions:**

- illegitimate objective;
- unnecessary restriction;
- hidden burden;
- absent alternatives;
- rights traded away;
- affected communities excluded.

**Architectural significance:** Wisdom asks not only whether an action can achieve an end, but whether this means of achieving it deserves use.

## Moral Uncertainty

Suppose plausible ethical frameworks are:

$$M_E = \{M_1, \ldots, M_K\}.$$

Let:

$$w_k(a)$$

be the evaluation of action $a$ under framework $M_k$.

Let:

$$q_k$$

be the confidence assigned to that framework.

If cross-framework values can be meaningfully normalized, one may form:

$$\bar{W}(a) = \sum_{k=1}^{K} q_k \widetilde{w}_k(a).$$

But such normalization is philosophically demanding.

One framework may regard a violation as categorically impermissible.

Another may treat the same outcome as compensable.

When scales are not comparable, a robust ethical set may be safer:

$$A_{eth} = \{a : h_{k,j}(a) \leq 0 \,\forall j \text{ under every sufficiently plausible } M_k\}.$$

### Equation Passport 23.11

***Judgment Under Moral Uncertainty***

$$\bar{W}(a) = \sum_{k} q_k \widetilde{w}_k(a)$$

when comparisons are justified, or:

$$A_{eth} = ¿ k\, A_{permitted}^{(k)}$$

when they are not.

**Status:** $[ARC/MOD/CONJ]$

**Ethical frameworks:** $M_k$

**Framework confidence:** $q_k$

**Framework-specific evaluation:** $w_k$

**Required declarations:**

- why frameworks were included;
- whether comparisons are meaningful;
- whether rights are compensable;
- whether one framework dominates the scale;
- what uncertainty remains.

**What moral uncertainty does not justify:** Moral indifference

**Architectural significance:** When certainty about ethics is unavailable, wisdom may seek actions that remain defensible across several serious moral perspectives.

## Dignity Is Not a Small Weight

Suppose stakeholder $i$ possesses dignity condition:

$$d_i(a) \geq \theta_i.$$

A weighted objective might permit severe violation of $d_i$ if benefits elsewhere are large enough.

The Architecture rejects this silent compensation where dignity is constitutional.

Define the dignity margin:

$$\mu_D(a) = \min_i \frac{d_i(a) - \theta_i}{s_i}.$$

A dignity-preserving action satisfies:

$$\mu_D(a) \geq 0.$$

UNESCO's bioethics declaration places human dignity and individual welfare before the sole interest of science or society and requires autonomy, consent, and protection of those unable to exercise autonomy fully.

### Equation Passport 23.12

*Dignity Floor*

$$\mu_D(a) = \min_i \frac{d_i(a) - \theta_i}{s_i}$$

**Status:** $[ARC/MOD]$

**Stakeholder standing:** $d_i$

**Minimum protected condition:** $\theta_i$

**Scale:** $s_i$

**Admissibility condition:**

$$\mu_D(a) \geq 0.$$

**What it prevents:** Aggregate benefit concealing the destruction of one person's fundamental standing

**What it does not provide:** A universal numerical measure of dignity

**Architectural significance:** Persons remain ends within the decision boundary rather than variables to be optimized away.

## Compassion Enlarges the Boundary

Let:

$$L_i(a, \theta)$$

be the loss borne by stakeholder $i$.

A self-enclosed decision considers only:

$$L_{self}.$$

A relational architecture considers:

$$L_{rel} = \begin{bmatrix} L_1 \\ \vdots \\ L_n \\ L_{future} \\ L_{ecological} \end{bmatrix}.$$

Compassion begins when another being's vulnerability is admitted as relevant before the decision is made.

Compassion does not mean:

- deciding for others without consent;
- denying conflict;
- removing accountability;
- accepting every demand;
- erasing boundaries.

Compassion enlarges who is seen.

Wisdom still must determine what responsible care requires.

Book I recognized compassion as one of coherence's highest mature expressions because shared reality, biological inheritance, vulnerability, meaning, and mortality connect human beings beyond abstraction.

## Responsibility

Responsibility is rarely one scalar.

For agent $i$, define a responsibility signature:

$$R_i = \begin{bmatrix} \kappa_i \\ f_i \\ c_i \\ o_i \\ b_i \\ r_i \end{bmatrix} [ARC/MOD]$$

where:

- $\kappa_i$ is causal contribution;
- $f_i$ is foreseeability;
- $c_i$ is effective control;
- $o_i$ is role obligation;
- $b_i$ is benefit received;
- $r_i$ is capacity and duty to repair.

A person can be causally involved without possessing full control.

A person can possess authority without direct physical involvement.

An institution can benefit from a harmful structure it did not originally create.

A present generation can inherit consequences without personally causing their origin.

Responsibility therefore includes more than blame.

It includes repair.

## Shared Causal Contribution

Let:

$$N$$

be a set of participants.

Let:

$$v(S)$$

represent the outcome produced by coalition $S \subseteq N$.

The Shapley value assigns participant $i$:

$$\phi_i(v) = \sum_{S \subseteq N \setminus \{i\}} \frac{|S|!(n-|S|-1)!}{n!}\left[v(S \cup \{i\}) - v(S)\right].$$

Shapley derived this cooperative-game allocation from axiomatic properties concerning contribution. It is useful for dividing modeled joint value or cost; it does not, by itself, determine moral blame, legal liability, intention, or duty.

## Equation Passport 23.13

*Shared Contribution*

$$\phi_i(v) = \sum_{S \subseteq N \setminus \{i\}} \frac{|S|!(n-|S|-1)!}{n!}[v(S \cup \{i\}) - v(S)]$$

**Status:** $[THM/MOD]$

**Participants:** $N$

**Coalition value:** $v(S)$

**Marginal contribution:** $v(S \cup \{i\}) - v(S)$

**Allocation:** $\phi_i$

**What it can support:**

- cost allocation;
- benefit sharing;
- contribution analysis;
- distributed credit.

**What it does not establish:**

- intention;
- culpability;
- knowledge;
- authority;
- coercion;
- duty to repair.

**Architectural significance:** Mathematics can help reveal distributed contribution while wisdom preserves the larger responsibility architecture.

# Responsibility Without Control

A patient may bear consequences without controlling the institution.

A child may inherit debt without participating in its creation.

A worker may implement policy without designing it.

A leader may authorize harm without performing it directly.

The responsibility architecture must distinguish:

Bearer of HarmCausal ContributorDecision AuthorityBeneficiary Repairing Agent.

These roles may belong to different people.

## Accountability

A wise decision should leave a recoverable record.

Define a decision record:

$$D_t = (A_t, E_t, M_t, P_t, V_t, H_t, a_t, \hat{Y}_t)$$

where:

- $A_t$is the considered alternative set;
- $E_t$is evidence;
- $M_t$is the model;
- $P_t$is the probability or scenario architecture;
- $V_t$is the value architecture;
- $H_t$is the harm and constraint architecture;
- $a_t$is the action;
- $\hat{Y}_t$is the predicted outcome.

After the outcome:

$$Y_t$$

appears, define discrepancy:

$$\delta_t = Y_t - \hat{Y}_t.$$

The accountability update is:

$$D_{t+1} = Revise(D_t, Y_t, \delta_t, C_t),$$

where $C_t$includes criticism, testimony, and review.

## Equation Passport 23.14

*Accountability Loop*

$$D_t = (A_t, E_t, M_t, P_t, V_t, H_t, a_t, \hat{Y}_t) \delta_t = Y_t - \hat{Y}_t$$

$$D_{t+1} = Revise\ (D_t, Y_t, \delta_t, C_t)$$

**Status:** $[ARC / MOD]$

**Purpose:** Preserve the reasons and assumptions behind action

**Accountability requires:**

- traceability;
- explanation;
- review;
- correction;
- remedy;
- retained institutional memory.

**Failure conditions:**

- undocumented alternatives;
- deleted evidence;
- post hoc justification;
- outcome concealment;
- absence of appeal;
- no responsible repair agent.

**Architectural significance:** Wisdom does not merely choose; it leaves the decision open to truth after the decision-maker's authority has passed.

# Accountability Is Not Punishment Alone

Accountability may include:

- explanation;
- acknowledgment;
- correction;
- restoration;
- compensation;
- institutional redesign;
- prevention.

Punishment without learning may preserve anger while leaving the harmful architecture unchanged.

Learning without responsibility may protect institutions from consequence.

Wisdom joins both.

## Stewardship

Book I declared:

Reality entrusted every generation with an inheritance.

The Earth.

Knowledge.

Mathematics.

Language.

Culture.

Science.

Children.

None originated entirely with us.

None belongs entirely to us.

Stewardship is therefore neither passive preservation nor unrestricted use.

It is temporary custodianship joined to improvement and transmission.

Let the inherited capital of generation $g$ be:

$$H_g = \begin{bmatrix} H_g^N \\ H_g^K \\ H_g^C \\ H_g^I \\ H_g^R \\ H_g^T \end{bmatrix}$$

where:

- $H_g^N$ is natural inheritance;
- $H_g^K$ is knowledge;
- $H_g^C$ is cultural inheritance;
- $H_g^I$ is institutional capacity;
- $H_g^R$ is relational trust;
- $H_g^T$ is technical capability.

Transmission follows:

$$H_{g+1} = T_g H_g + G_g - D_g,$$

where:

- $T_g$ is the transmission architecture;
- $G_g$ is new contribution;
- $D_g$ is depletion, destruction, corruption, or loss.

## Equation Passport 23.15

***Stewardship of Inheritance***

$$H_{g+1} = T_g H_g + G_g - D_g$$

**Status:** $[ARC/MOD]$

**Inherited capital:** $H_g$

**Transmission:** $T_g$

**New contribution:** $G_g$

**Depletion:** $D_g$

**Stewardship questions:**

- what was received;
- what was consumed;
- what was strengthened;
- what was corrupted;
- what remains available;
- what future freedom was preserved.

**Failure condition:** One form of capital is increased by irreversibly destroying another without disclosure.

**Architectural significance:** Wisdom measures success partly by the condition in which the inheritance is handed onward.

## Stewardship Is Not Freezing the Past

Faithful stewardship may require:

- repair;
- translation;
- correction;
- reinterpretation;
- release of obsolete practice;
- creation of new capacity.

A language survives by being spoken and extended.

Science survives by correcting itself.

Culture survives by remaining transmissible and capable of responding to reality.

An institution survives coherently not by preserving every rule, but by preserving the legitimate purpose for which the rules existed.

Stewardship Preserves Fidelity, Not Every Form.

## Intergenerational Judgment

Let generation index be:

$$g=0,1,\ldots,G.$$

A discounted social objective may be:

$$W=\sum_{g=0}^{G}\delta^{g}U_{g},0<\delta\leq 1.$$

Ramsey's mathematical theory of saving formalized an early intertemporal question concerning how present consumption and future welfare should be related. Such models make temporal tradeoffs explicit; they do not determine the morally correct discounting of future persons.

A constitutional alternative preserves generation-specific floors:

$$c_{g,j}\geq\theta_{j}\forall g,j.$$

Let:

$$\Omega_{g}$$

be the set of meaningful choices available to generation $g$.

An option-preserving stewardship condition is:

$$\mu\left(\Omega_{g+1}\right)\geq\underline{\Omega}.$$

UNESCO's future-generations declaration asks present generations to preserve human dignity, ecosystems, natural resources, cultural heritage, peace, freedom of choice, education, and the consequences of major projects for those who will inherit them.

### Equation Passport 23.16

*Intergenerational Viability*

$$W=\sum_{g=0}^{G}\delta^{g}U_{g}$$

subject to:

$$c_{g,j}\geq\theta_{j}$$

and:

$$\mu\left(\Omega_{g+1}\right) \geq \underline{\Omega}.$$

**Status:** $[MOD/ARC]$

**Generation welfare:** $U_g$

**Discount factor:** $\delta$

**Capability floor:** $\theta_j$

**Future option set:** $\Omega_{g+1}$

**Required declarations:**

- why future utility is discounted;
- who represents unborn persons;
- which losses are irreversible;
- which capabilities must remain;
- what uncertainty surrounds distant futures.

**What the scalar objective cannot settle:** Equal moral standing across time

**Architectural significance:** Wisdom does not merely transmit resources; it transmits the possibility of meaningful choice.

## Future Generations Cannot Consent

Future persons cannot:

- vote in present elections;
- refuse present contracts;
- challenge present models;
- inspect present risk assessments;
- withdraw consent.

This absence does not remove them from the causal architecture.

It increases the stewardship burden upon those who act now.

A civilization becomes mature when political absence does not become moral invisibility.

## Discounting and Moral Standing

Financial discounting may represent:

- opportunity cost;
- investment return;
- uncertainty;
- impatience.

It must not silently imply:

A person matters less because they live later.

Monetary discounting and moral discounting are not identical.

Long-term judgment should distinguish:

$$\text{Present Value of Money}$$

from:

$$\text{Standing of Future Life}.$$

## Irreversible Inheritance

UNESCO's declaration states that present generations should not bequeath an Earth irreversibly damaged by human activity and should preserve natural resources, ecosystems, cultural diversity, and heritage for those who follow.

The Architecture generalizes the inheritance question carefully:

**What present benefit becomes illegitimate when purchased through the permanent removal of another generation's basic possibilities?**

## Governing the Commons

Let shared resource stock be:

$$x_t.$$

Let natural renewal be:

$$g(x_t).$$

Let total extraction be:

$$h_t = \sum_{i=1}^{n} h_{i,t}.$$

Let restoration be:

$$r_t.$$

Then:

$$x_{t+1} = x_t + g(x_t) - h_t + r_t.$$

A viability condition is:

$$x_t \geq x_{min} \forall t.$$

Individual short-term incentives may encourage extraction beyond the long-term viable level.

Institutional rules can alter:

- access;
- monitoring;
- contribution;
- sanction;
- conflict resolution;
- local adaptation.

Elinor Ostrom's work showed that common-pool resources need not be governed only through centralized control or private ownership; communities can develop durable, context-dependent, often polycentric institutions for collective resource governance.

## Equation Passport 23.17

***Commons Stewardship***

$$x_{t+1} = x_t + g(x_t) - \sum_i h_{i,t} + r_t$$

subject to:

$$x_t \geq x_{min}.$$

**Status:** $[MOD]$

**Resource stock:** $x_t$

**Regeneration:** $g(x_t)$

**Extraction:** $h_{i,t}$

**Restoration:** $r_t$

**Governance variables may include:**

- membership;
- monitoring;
- contribution rules;
- sanctions;
- dispute resolution;
- nested institutions;
- local revision.

**Failure conditions:**

- extraction hidden;
- users excluded from rule formation;
- benefits privatized;
- losses externalized;
- future regeneration overestimated.

**Architectural significance:** Stewardship becomes civilization when shared inheritance is governed through relationships strong enough to preserve both use and renewal.

## Shared Ownership Does Not Mean Ownerless

When everyone depends upon a resource but no one accepts responsibility, the commons becomes administratively invisible.

When one actor claims exclusive control, the commons may become extractive.

Wise governance joins:

- access;
- boundary;
- responsibility;
- monitoring;
- participation;
- repair.

## Adaptive Stewardship

A fixed rule may be coherent under one climate, technology, population, or institution and incoherent under another.

Let policy be:

$$\pi_t.$$

Let monitoring produce:

$$y_t.$$

Let threshold discrepancy be:

$$e_t = y_t - y_t^{expected}.$$

An adaptive stewardship rule is:

$$\pi_{t+1} = U(\pi_t, e_t, E_t, V_t).$$

Stewardship does not only preserve the resource.

It preserves the institution's capacity to notice when its own rules are failing.

## Wisdom and Collective Deliberation

No single person ordinarily possesses:

- all relevant facts;
- every lived perspective;
- every discipline;
- every consequence;
- complete moral insight.

Collective wisdom therefore depends upon:

- distributed knowledge;
- independent evidence;
- affected-person testimony;
- disagreement;

- public reasoning;
- correction;
- institutional memory.

But more participants do not automatically create wiser judgment.

If all participants inherit one mistaken source:

$$\rho_{error} \approx 1,$$

their agreement may magnify confidence without reducing error.

A wise deliberative architecture must preserve both:

$$\text{Coordination}$$

and:

$$\text{Corrective Difference}.$$

## Expertise and Experience

Technical experts may understand:

- mechanisms;
- probability;
- engineering;
- clinical evidence;
- model uncertainty.

Affected persons may understand:

- lived burden;
- inaccessible alternatives;
- cultural meaning;
- trust;
- practical feasibility;
- harms absent from the model.

Neither knowledge source should erase the other.

The complete decision may require:

$$\text{Technical Expertise} + \text{Lived Expertise} + \text{Ethical Accountability}.$$

# Wisdom and Time

A decision can appear wise over one month and disastrous over one generation.

Let consequence at horizon $T$ be:

$$Y_T(a).$$

Define a horizon-indexed evaluation:

$$J_T(a) = E\left[L\left(Y_T(a)\right)\right].$$

An action may satisfy:

$$J_{T_1}(a) < J_{T_1}(b)$$

for short horizon $T_1$, while:

$$J_{T_2}(a) > J_{T_2}(b)$$

for longer horizon $T_2$.

Wisdom declares the horizon.

Short-term visibility should not become permanent authority over long-term consequence.

# Wisdom and Scale

An action may benefit:

- one person;
- one institution;
- one nation;
- one generation.

It may harm:

- another community;
- another ecosystem;
- another scale;

- another time.

Let:

$$L_{l,g,i}(a)$$

be loss at scale $l$, generation $g$, and stakeholder $i$.

A wise assessment preserves the tensor:

$$L(a)=[L_{l,g,i}(a)]$$

long enough to identify where burdens are being hidden.

Aggregation may follow.

Visibility must come first.

## The Principle of Living Coherence

Book I's final constitutional principle declared:

**Coherence reaches its highest expression when organized understanding continually strengthens life, dignity, responsibility, and future inheritance.**

Book II can represent this as a living-coherence consequence vector:

$$\Delta Z(a)=\begin{bmatrix} \Delta L_f(a) \\ \Delta D_g(a) \\ \Delta R_s(a) \\ \Delta I_h(a) \end{bmatrix} [ARC/MOD]$$

where:

- $\Delta L_f$ is change in life-supporting capacity;
- $\Delta D_g$ is change in dignity preservation;
- $\Delta R_s$ is change in responsible capability;
- $\Delta I_h$ is change in future inheritance.

A wise action should not be judged by the sum alone.

One dimension may be noncompensable.

## Equation Passport 23.18

*Living-Coherence Test*

$$\Delta Z(a) = (\Delta, L_f, \Delta, D_g \Delta R_s \Delta I_h)^T$$

**Status:** $[ARC/MOD]$

**Life:** $\Delta L_f$

**Dignity:** $\Delta D_g$

**Responsibility:** $\Delta R_s$

**Inheritance:** $\Delta I_h$

**Constitutional use:**

- preserve dimensions separately;
- identify tradeoffs;
- prohibit dignity violations;
- include future consequences;
- retain uncertainty.

**Non-claim:** No universal numerical scale of moral organization is established.

**Architectural significance:** Wisdom is not merely coherence preserved; it is coherence directed toward life, dignity, responsibility, and inheritance.

# The Wisdom Signature

A mature wisdom assessment should remain multidimensional.

Define:

$$Q_W = \begin{bmatrix} K_f \\ U_h \\ C_x \\ V_p \\ P_r \\ C_p \\ A_c \\ F_i \\ R_v \\ S_t \end{bmatrix} [ARC/MOD]$$

where:

- $K_f$ is knowledge fidelity;
- $U_h$ is humility before uncertainty;
- $C_x$ is contextual understanding;
- $V_p$ is recognition of plural values;
- $P_r$ is proportionality;
- $C_p$ is compassion;
- $A_c$ is accountability;
- $F_i$ is future-inheritance preservation;
- $R_v$ is reversibility or option preservation;
- $S_t$ is stewardship.

## Equation Passport 23.19

*Wisdom-Quality Signature*

$$Q_W = (K_f, U_h, C_x, V_p, P_r, C_p, A_c, F_i, R_v, S_t)^T$$

**Status:** $[ARC/MOD]$

**Purpose:** Preserve the distinguishable burdens carried by wise judgment

**Possible failure patterns:**

- strong knowledge, weak compassion;
- strong intention, weak evidence;

- strong present benefit, weak future inheritance;
- strong accountability, weak reversibility;
- strong consensus, weak dignity.

**What it prevents:** One impressive virtue concealing a decisive failure elsewhere

**Architectural significance:** Wisdom is a configuration of capacities rather than one score of superiority.

## Operational Descendant of Book I's Wisdom Recognition

Book I declared:

$$\text{Wisdom} = \text{coherent understanding faithfully lived}. [ARC]$$

An operational descendant is:

$$W_t = F(K_t, M e_t, C_t, J_t, P_t, A_t, S_t F_t) [ARC/MOD]$$

where:

- $K_t$ is organized knowledge;
- $M e_t$ is meaningful orientation;
- $C_t$ is coherence;
- $J_t$ is contextual judgment;
- $P_t$ is compassion and proportionality;
- $A_t$ is accountability;
- $S_t$ is stewardship;
- $F_t$ is future responsibility.

The quality of lived participation may be assessed through:

$$\Delta W_t = W_{t+1} - W_t,$$

but the number is not the wisdom.

The evidence lies in:

- what was understood;
- how uncertainty was handled;
- who was protected;
- what was repaired;
- what remained teachable;
- what was handed forward.

## Equation Passport 23.20

*Operational Wisdom Architecture*

$$W_t = F\left(K_t, M\,e_t, C_t, J_t, P_t, A_t, S_t\, F_t\right)$$

**Status:** $[ARC/MOD]$

**Knowledge:** $K_t$

**Meaning:** $M\,e_t$

**Coherence:** $C_t$

**Judgment:** $J_t$

**Compassion and proportionality:** $P_t$

**Accountability:** $A_t$

**Stewardship:** $S_t$

**Future responsibility:** $F_t$

**What it adds to Book I:** Explicit decision, uncertainty, accountability, and future-inheritance dimensions

**What it does not claim:** One universal empirical function of wisdom

**Architectural significance:** Understanding becomes wisdom only through the quality of the participation it produces.

# Wisdom and Medicine

Medicine reveals the final chamber with unusual clarity.

A clinical decision may involve:

- evidence from populations;
- uncertainty concerning one patient;
- possible benefit;
- possible harm;

- patient values;
- urgency;
- reversibility;
- consent;
- quality of life;
- family and social context.

The statistically highest expected benefit may not be the patient's preferred path.

The patient's preference may be insufficiently informed.

The clinician may know mechanism but not the patient's lived priorities.

A wise clinical architecture therefore joins:

$$\text{Evidence} + \text{Expertise} + \text{Patient Values} + \text{Consent} + \text{Proportionality}.$$

UNESCO's bioethics framework places dignity, individual welfare, benefit and harm, autonomy, consent, equality, justice, solidarity, and protection of vulnerable persons within scientific and medical judgment.

## Wisdom and Science

Science can tell us:

- what appears possible;
- what mechanisms operate;
- what evidence supports;
- what uncertainty remains;
- what consequence may follow.

Science cannot, by itself, settle every question of use.

The same discovery may support:

- healing;
- surveillance;
- liberation;
- exploitation;
- education;
- control.

The Constitution remains:

$$\text{Scientific Possibility} \not\Rightarrow \text{Ethical Permission}.$$

Wisdom does not restrain science by diminishing truth.

It protects science from becoming detached from responsibility.

## Wisdom and Technology

A technology should not be evaluated only by:

- accuracy;
- speed;
- scale;
- profit;
- adoption.

Its decision boundary should include:

- affected persons;
- dignity;
- consent;
- explainability;
- exclusion;
- power concentration;
- environmental burden;
- future lock-in;
- means of appeal;
- reversibility.

UNESCO's AI recommendation explicitly connects technical development with human dignity, human rights, environmental and ecosystem protection, proportionality, impact assessment, accountability, traceability, plural stakeholder participation, and the interests of present and future generations.

## Machine Optimization Is Not Wisdom

An artificial system may:

- estimate probabilities;
- optimize policies;
- model scenarios;
- calculate CVaR;
- identify Pareto frontiers;

- recommend actions;
- generate moral language.

These are powerful functions.

They do not establish:

- lived concern;
- compassion;
- conscience;
- moral standing;
- subjective responsibility;
- human-like wisdom.

A machine may support wise judgment.

It should not silently inherit authority over the values, dignity, or future of persons whose lives it does not experience.

## Moral Authority Cannot Be Delegated Invisibly

When an automated system influences:

- health;
- employment;
- liberty;
- welfare;
- education;
- credit;
- public speech,

the institution remains responsible for:

- the objective;
- the data;
- the model;
- the threshold;
- the appeal process;
- the resulting harm.

The phrase:

The algorithm decided

is not an adequate accountability structure.

## Dual-Aspect Wisdom

Let:

$$W_o(t)$$

represent exterior participatory wisdom.

Let:

$$W_i(t)$$

represent modeled interior wisdom.

Let:

$$W_b(t)$$

represent wisdom at the boundary joining intention, action, communication, and consequence.

A dual-aspect wisdom state is:

$$W_{DA}=\begin{bmatrix} W_o \\ W_i \\ W_b \end{bmatrix}.$$

Exterior wisdom may include:

- proportionate action;
- measurable protection;
- responsible allocation;
- repair;
- stewardship.

Interior modeled wisdom may include:

- humility;
- compassion;

- integrity;
- willingness to be corrected;
- alignment between value and intention.

Boundary wisdom may include:

- consent;
- truthful communication;
- accountability;
- relationship between intention and actual consequence.

The dynamics may be:

$$\frac{dW_{DA}}{dt} = G_W - L_W [ARC/MOD/CONJ].$$

## Equation Passport 23.21

*Dual-Aspect Wisdom Architecture*

$$W_{DA} = (W_o, W_i, W_b)^T \dot{W}_{DA} = G_W - L_W$$

**Status:** $[ARC/MOD/CONJ]$

**Exterior participation:** $W_o$

**Interior modeled integrity:** $W_i$

**Interface:** $W_b$

**Required empirical work:**

- operational definitions;
- longitudinal behavior;
- report and observation;
- consequence assessment;
- comparison with competing explanations.

**Non-claims:**

- no separate wisdom substance;
- no physical wisdom field;
- no universal neural wisdom score;
- no claim that good intention equals wise consequence.

**Architectural significance:** Wisdom must become visible in participation while remaining rooted in an interior capacity for integrity and correction.

## Good Intention Is Not Enough

A person may intend care and cause harm.

An institution may intend efficiency and produce exclusion.

A movement may intend justice and reproduce domination.

Wisdom therefore requires:

$$\text{Intention} + \text{Competence} + \text{Consequence} + \text{Correction}.$$

Compassion without knowledge can become dangerous.

Knowledge without compassion can become cruel.

## The Spirit Field and Wisdom

The Spirit Field remains:

$$S_f[SYM/ARC]$$

as symbolic language for organized interior coherence.

A symbolic wisdom state may be:

$$S_f^W = O(U, G_r, H, R_s, F_g, W_n, P_u, C_p, S_t)[SYM/ARC]$$

where:

- $U$ is understanding;
- $G_r$ is gratitude;
- $H$ is hope;
- $R_s$ is responsibility;
- $F_g$ is forgiveness;
- $W_n$ is wonder;
- $P_u$ is purpose;
- $C_p$ is compassion;
- $S_t$ is stewardship.

This may support:

- reflection;
- teaching;
- spiritual practice;
- ethical examination;
- relational repair;
- ceremony;
- grief;
- future responsibility.

It does not establish:

- a physical field;
- a new energy;
- a measurable force;
- a quantum mechanism;
- a clinical diagnosis;
- a supernatural decision procedure.

## Gratitude and Stewardship

Gratitude recognizes:

I received before I contributed.

Stewardship answers:

Therefore I must transmit faithfully.

Gratitude without stewardship becomes sentiment.

Stewardship without gratitude may become control.

## Hope and Judgment

Hope permits future possibility to influence present action.

False hope denies evidence.

Despair converts uncertainty into certainty of failure.

Wise hope can be written symbolically as:

$$H_W = \text{Possibility} + \text{Value} + \text{Agency} + \text{Truthfulness}.$$

Hope does not require high probability.

It requires that the possibility remain worthy of faithful participation.

## Forgiveness and Accountability

Forgiveness does not require:

- erasing history;
- restoring unsafe access;
- ending accountability;
- removing boundaries;
- denying injury.

Forgiveness may change how the past governs the future while preserving truth.

Wisdom protects both:

$$\text{Release}$$

and:

$$\text{Record}.$$

## Wonder and Knowledge

Wonder need not disappear when mechanism becomes known.

Scientific understanding can enlarge wonder by revealing:

- timescale;
- relation;
- complexity;
- inheritance;
- improbability;
- lawful structure.

Wisdom does not use mystery to protect ignorance.

It allows knowledge to deepen reverence.

## The Bridge Species as Moral Participant

The Bridge Species can represent:

- persons not presently visible;
- future generations;
- distant ecosystems;
- alternative histories;
- consequences beyond one lifetime;
- the possibility that its own values are wrong.

This capacity enlarges responsibility.

The Bridge Species becomes mature not merely when it predicts the future, but when it permits the future to place a claim upon the present.

## Civilization

Book I rejected wealth, technology, military power, and monuments as sufficient measures of civilization.

It located civilization in the quality of understanding transmitted across generations:

- education deepening;
- science remaining honest;
- mathematics remaining rigorous;
- justice remaining trustworthy;
- compassion remaining visible;
- wisdom reorganizing knowledge toward life.

A civilizational state may be represented as:

$$C_{civ} = \begin{bmatrix} C_{truth} \\ C_{education} \\ C_{justice} \\ C_{health} \\ C_{compassion} \\ C_{ecology} \\ C_{inheritance} \end{bmatrix}.$$

A civilization cannot claim maturity because one component is large.

Its constitutional viability requires:

$$C_i \geq \theta_i \forall i.$$

## Civilization Is Continually Taught

Buildings can remain after the civilization that understood them has disappeared.

Books can remain unread.

Equations can remain uncopied.

Archives can remain inaccessible.

Civilization continues only when living persons inherit:

- methods;
- memory;
- judgment;
- compassion;
- responsibility.

Book I called these persons **the living stones**.

Teachers.

Scientists.

Physicians.

Engineers.

Parents.

Children who learn to love reality.

Human beings faithfully shaped by reality become the enduring architecture through which civilization continues.

## Common Failures of Wisdom Architecture

### Wisdom–Intelligence Collapse

High cognitive ability is treated as wise judgment.

### Wisdom–Age Collapse

Age is treated as automatic wisdom.

### Wisdom–Authority Collapse

Institutional rank is treated as evidence of wisdom.

### Knowledge–Wisdom Collapse

Correct explanation is treated as responsible participation.

### Meaning–Wisdom Collapse

Strong purpose is treated as wise action.

### Coherence–Goodness Collapse

A system's capacity to preserve itself is treated as moral legitimacy.

### Outcome Bias

A lucky outcome is used to justify a reckless decision.

## Hindsight Bias

Knowledge available after the event is projected backward into the earlier decision.

## Optimization Worship

A mathematically optimal result is treated as wise without examining the objective.

## Objective Concealment

Values enter the model without being declared.

## Constraint Concealment

The model hides what was treated as inviolable.

## Alternative-Set Concealment

The chosen action appears necessary because other alternatives were never considered.

## Formal-Choice Illusion

An option is counted as available even though cost, coercion, disability, or information makes it inaccessible.

## Expected-Value Absolutism

Average consequence conceals severe tail harm.

## Worst-Case Absolutism

One extreme scenario dominates every decision regardless of plausibility.

## Regret Absolutism

Avoidance of future self-reproach replaces responsibility to actual persons.

## Robustness–Wisdom Collapse

Performance across scenarios is treated as ethical legitimacy.

### Satisficing Complacency

A low aspiration threshold protects mediocrity or injustice.

### Information Worship

More data are collected despite no decision relevance.

### Information Delay Blindness

Research continues while urgent harm grows.

### Information-Harm Erasure

Privacy, consent, and burden of data collection are omitted.

### Certainty Requirement

Action is postponed until impossible certainty appears.

### Uncertainty Excuse

Absence of certainty is used to ignore credible danger.

### Precautionary Paralysis

Every unfamiliar action is prohibited.

### Precautionary Opportunism

Precaution is invoked selectively to protect existing power.

### Tail-Risk Erasure

Catastrophic outcomes are averaged away.

### Probability-Model Concealment

Precise probabilities are presented despite deep ambiguity.

## Unknown-Unknown Erasure

The modeled scenario set is treated as complete reality.

## Reversibility Erasure

Future lock-in is omitted from evaluation.

## Option-Value Erasure

A decision closes future possibilities without accounting for what was lost.

## Status-Quo Neutrality

Inaction is treated as consequence-free.

## Proportionality Erasure

A useful intervention is assumed justified regardless of burden.

## Effectiveness–Permission Collapse

The fact that a method works is treated as sufficient moral reason to use it.

## Less-Restrictive-Alternative Erasure

A more harmful method is chosen without comparing safer means.

## Dignity as a Weight

Fundamental human standing becomes compensable through aggregate benefit.

## Consent Formalism

A signature is treated as genuine consent despite misinformation or coercion.

## Autonomy–Abandonment Collapse

Respect for choice is used to withdraw needed support.

### Paternalism

Care becomes control over another person's life.

### Preference–Welfare Collapse

Expressed preference is treated as complete well-being.

### Behavior–Value Collapse

Observed action is treated as unique evidence of true value.

### Moral-Certainty Theater

Confidence is performed where ethical conflict remains unresolved.

### Moral-Uncertainty Indifference

Plural ethical perspectives are used as an excuse to choose arbitrarily.

### Framework Normalization Concealment

Incommensurable moral theories are forced onto one scale without justification.

### Stakeholder Erasure

Affected persons remain outside the model.

### Future-Generation Erasure

Only presently represented preferences enter the decision.

### Ecological Externality Erasure

Nonhuman systems absorb losses excluded from the objective.

### Boundary Manipulation

The decision-maker narrows the system until exported harms disappear.

## Contribution–Blame Collapse

A causal allocation formula is treated as complete moral responsibility.

## Authority–Control Confusion

A person with title is assumed to possess actual control, or actual control is hidden behind title.

## Benefit Erasure

Those who profit from a harmful system are omitted from responsibility.

## Repair-Duty Erasure

Responsibility ends with acknowledgment rather than correction.

## Accountability–Punishment Collapse

Punishment replaces learning, repair, and prevention.

## Accountability Theater

Records exist but no one can appeal, investigate, or obtain remedy.

## Post-Hoc Rationalization

Reasons are invented after the outcome.

## Stewardship–Ownership Collapse

Temporary custodians act as though inheritance belongs exclusively to them.

## Preservation Worship

Every inherited practice is protected because it is old.

## Innovation Worship

Every inherited practice is discarded because it is old.

### Heritage Fabrication

Desired ancestral continuity replaces evidence.

### Cultural Freezing

A living inheritance is treated as though faithfulness requires immobility.

### Discounting–Moral-Standing Collapse

Financial discounting becomes reduced concern for future persons.

### Intergenerational Extraction

Present benefit is purchased through irreversible future deprivation.

### Commons Fatalism

Shared resources are assumed incapable of organized governance.

### Centralization Absolutism

One authority is assumed universally superior.

### Privatization Absolutism

Exclusive ownership is assumed universally superior.

### Participation Theater

Stakeholders are consulted after the decisive architecture is fixed.

### Consensus–Wisdom Collapse

Agreement is treated as evidence that the decision is just or true.

### Dissent Romanticism

Opposition alone is treated as wisdom.

## Expertise Export

Authority in one domain is extended beyond competence.

## Lived-Experience Absolutism

Personal experience replaces population evidence and causal inquiry.

## Technical-Expertise Absolutism

Formal expertise erases lived consequence.

## Compassion Without Competence

Careful intention substitutes for effective action.

## Competence Without Compassion

Technical success becomes detached from dignity.

## Justice Without Reality

Moral aspiration ignores factual consequence.

## Reality Without Justice

Measurement is used to evade normative responsibility.

## AI Authority Transfer

Institutions hide moral decisions behind automated outputs.

## Machine-Wisdom Anthropomorphism

Optimization and moral language are treated as lived wisdom.

## Human-Oversight Theater

A human nominally remains involved but lacks time, authority, or understanding to intervene.

### Spirit-Field Physicalization

Symbolic wisdom is converted into an unsupported physical mechanism.

## The Wisdom Constitution

Before any wisdom claim enters Book II, it must answer:

1. What decision is being made?
2. Who possesses authority to decide?
3. Who possesses actual control?
4. Who is affected?
5. Who is absent?
6. Who will inherit the consequence?
7. What evidence is available?
8. What evidence is unavailable?
9. What evidence may arrive later?
10. What information could change the decision?
11. What information cannot change it?
12. What is the cost of obtaining more information?
13. What is the cost of waiting?
14. What is the cost of acting now?
15. What is the state space?
16. What uncertainties are represented?
17. Which uncertainties are aleatoric?
18. Which are epistemic?
19. Which are ambiguous?
20. Which are deeply uncertain?
21. What possibilities lie outside the model?
22. What probability distribution is used?
23. Who selected it?
24. How was it calibrated?
25. What model family is used?
26. What alternative models remain?
27. What assumptions connect action to consequence?
28. What causal pathways are known?
29. What causal pathways remain speculative?
30. What action set is considered?
31. What actions were excluded?
32. Why were they excluded?
33. Are the alternatives genuinely available?
34. Are participants informed?
35. Are participants capable of refusing?
36. What happens if they refuse?
37. Is consent voluntary?
38. Is consent sufficiently informed?
39. Can consent later be withdrawn?
40. What decision rule is used?

41. Is expected loss minimized?
42. Is worst-case loss minimized?
43. Is regret minimized?
44. Is robustness sought?
45. Is satisficing used?
46. What aspiration level is used?
47. Is the aspiration level ethically sufficient?
48. What objective function is optimized?
49. Who selected the objective?
50. What does the objective omit?
51. Are several values present?
52. Are they preserved separately?
53. Are they forced into one scalar?
54. What weights are used?
55. Who selected the weights?
56. Are the units commensurable?
57. What constraints are noncompensable?
58. Is dignity treated as a constraint?
59. Are rights treated as constraints?
60. Is consent treated as a constraint?
61. Is justice treated as a constraint?
62. Is future viability treated as a constraint?
63. What benefits are expected?
64. Who receives the benefits?
65. What harms are expected?
66. Who bears the harms?
67. Are indirect harms included?
68. Are cumulative harms included?
69. Are ecological harms included?
70. Are future harms included?
71. Are distributional effects visible?
72. Is average loss concealing the worst affected person?
73. Is tail risk measured?
74. What confidence level is used?
75. Is CVaR or another tail measure used?
76. What catastrophe threshold is used?
77. Who defines catastrophe?
78. Is the probability of catastrophe known?
79. Is catastrophic harm compensable?
80. Is precaution invoked?
81. What credible harm pathway supports precaution?
82. Is the harm serious?
83. Is it irreversible?
84. Is uncertainty being used to justify delay?
85. Is uncertainty being used to justify prohibition?
86. Is the proposed precaution proportionate?
87. What harms arise from precaution itself?
88. Is the status quo treated as neutral?
89. Is the decision reversible?
90. What is the return path?
91. What is the return cost?

92. What future options are closed?
93. What future options are preserved?
94. Is option value measured?
95. Has irreversibility been earned?
96. Is the aim legitimate?
97. Is the intervention effective?
98. Is it necessary?
99. Is a less restrictive alternative available?
100. Is the burden proportionate?
101. Are rights preserved?
102. Is an appeal process available?
103. Is explanation available?
104. Is correction available?
105. Is remedy available?
106. Is moral uncertainty present?
107. Which ethical frameworks are plausible?
108. Can their evaluations be compared?
109. Are they being normalized arbitrarily?
110. Which actions remain permissible across them?
111. Does one framework identify a categorical violation?
112. Is the violation being averaged away?
113. What dignity floor applies?
114. Who selected the dignity indicators?
115. Can dignity be measured adequately?
116. What must remain outside numerical compensation?
117. Is compassion represented?
118. Does compassion preserve autonomy?
119. Does compassion become paternalism?
120. Is another person's vulnerability truly included?
121. Are nonhuman lives represented?
122. Are future generations represented?
123. Who speaks for them?
124. What causal contribution did each participant make?
125. What did each participant know?
126. What could each participant foresee?
127. What control did each participant possess?
128. What role obligation did each participant hold?
129. Who benefited?
130. Who can repair?
131. Is a Shapley or other allocation method used?
132. Is mathematical contribution being confused with blame?
133. Is responsibility distributed?
134. Is responsibility diluted until no one remains accountable?
135. Is the decision record preserved?
136. Were alternatives recorded?
137. Was evidence recorded?
138. Were assumptions recorded?
139. Were values recorded?
140. Were dissenting views recorded?
141. Was the predicted outcome recorded?
142. Was the actual outcome recorded?

143. Was discrepancy measured?
144. Was the model revised?
145. Was the institution revised?
146. Were harmed persons heard?
147. Was repair provided?
148. What inheritance was received?
149. What inheritance was consumed?
150. What inheritance was strengthened?
151. What inheritance was lost?
152. What inheritance was corrupted?
153. What will the next generation receive?
154. Will future generations retain meaningful choice?
155. What discount factor is used?
156. Why is it used?
157. Is money being discounted?
158. Is moral standing being discounted?
159. What capability floor is preserved?
160. What natural resources remain?
161. What cultural heritage remains?
162. What knowledge remains?
163. What institutional trust remains?
164. What educational capacity remains?
165. What peace and security remain?
166. Is a commons involved?
167. Who has access?
168. Who monitors use?
169. Who creates the rules?
170. Who may revise the rules?
171. Who resolves conflict?
172. Who restores the resource?
173. Are benefits private while losses are shared?
174. Is stewardship adaptive?
175. What monitoring triggers revision?
176. Does the policy learn?
177. Is technical expertise represented?
178. Is lived expertise represented?
179. Are errors independent?
180. Is consensus genuine?
181. Can dissent be expressed safely?
182. Is dissent evidence-based?
183. Are vulnerable participants included?
184. Is the time horizon explicit?
185. Is the spatial scale explicit?
186. Are local benefits damaging global viability?
187. Are present benefits damaging future viability?
188. Is wisdom being inferred from age?
189. Is wisdom being inferred from authority?
190. Is wisdom being inferred from confidence?
191. Is wisdom being inferred from eloquence?
192. Is wisdom being inferred from one good outcome?
193. Is machine optimization involved?

194. What objective was given to the machine?
195. What data shaped it?
196. What values were omitted?
197. Can a human actually intervene?
198. Who remains accountable?
199. Is dual-aspect wisdom being claimed?
200. What exterior participation supports it?
201. What interior evidence supports it?
202. What relationship joins intention and consequence?
203. Is the Spirit Field invoked?
204. Is its symbolic status explicit?
205. Is any physical mechanism being claimed?
206. What evidence would be required for such a claim?
207. Does the decision strengthen life?
208. Does it preserve dignity?
209. Does it enlarge responsibility?
210. Does it protect future inheritance?
211. Does it remain open to correction?
212. What would require reversal?
213. What would require apology?
214. What would require repair?
215. What would require institutional redesign?
216. What remains unknown?
217. What does the wisdom claim not establish?

A wisdom claim unable to answer these questions may remain useful as:

$$[ARC], [MOD], [SYM], \text{or } [CONJ].$$

Its status must remain visible.

## The Architect's Reflection

For many years, I believed wisdom would appear when enough knowledge had been gathered.

One more book.

One more theory.

One more equation.

One more discovery.

Reality gradually taught another lesson.

Knowledge can increase while compassion remains unchanged.

Technology can increase while dignity declines.

Power can increase while responsibility disappears.

A person can know exactly what is happening and still fail to do what the moment requires.

Wisdom did not appear merely when the answer became clearer.

It appeared when another person's life became impossible to treat as an abstraction.

When the question was no longer only:

What is true?

But also:

What does this truth now require of me?

Medicine taught me that a life cannot be reduced to a diagnosis.

Science taught me that certainty must remain proportionate to evidence.

History taught me that every generation inherits consequences it did not personally create.

Mathematics taught me that an elegant objective can conceal what was excluded from it.

Ordinary people taught me that wisdom often arrives quietly.

In patience.

In sacrifice.

In correction.

In forgiveness without forgetting.

In boundaries without resentment.

In remaining present when no equation can remove another person's pain.

Perhaps wisdom is not the possession of the final answer.

Perhaps it is the capacity to remain faithful when no answer removes responsibility.

The Cathedral was never built so that one generation could admire its own intelligence.

It was built so that people could enter.

So that truth could be protected.

So that dignity could remain visible.

So that another generation might begin beyond our mistakes.

Book I gave the Cathedral its oath.

Book II has measured its arches.

But no measurement explains why the Cathedral matters.

It matters because lives enter it.

Because children will inherit it.

Because truth needs guardians.

Because knowledge needs conscience.

Because coherence needs compassion.

Because civilization does not continue automatically.

Perhaps wisdom is the moment the builder realizes:

**The Cathedral is not mine.**

**The inheritance is not mine.**

**The future is not mine.**

**Yet each has been entrusted, for a time, to my participation.**

## Quiet Reflection

Pause before leaving the final chamber.

Think of one decision now before you.

Do not begin by asking which choice feels easiest.

Ask:

What do I know?

What do I only believe?

What remains uncertain?

Who will be affected?

Who is not present to speak?

What harm may be irreversible?

What future option may be closed?

What additional information would truly change the decision?

What would delay cost?

What value is hidden inside the objective?

What dignity must not be traded?

What responsibility belongs to you?

What responsibility belongs to others?

What inheritance did you receive?

What inheritance will your action leave?

Now think of one person whose wisdom shaped your life.

Perhaps they possessed no title.

Perhaps they never wrote a book.

Perhaps they simply knew when to speak.

When to listen.

When to protect.

When to forgive.

When to draw a boundary.

When to admit error.

When to remain.

Ask:

What did that person preserve within you?

What are you now carrying because they participated faithfully?

Remain with one final question:

**What does reality presently entrust to my judgment—and what kind of ancestor will this decision make me?**

## Constitutional Declaration of Wisdom Integrity

We will distinguish wisdom from intelligence.

We will distinguish wisdom from age.

We will distinguish wisdom from authority.

We will distinguish wisdom from confidence.

We will distinguish wisdom from eloquence.

We will distinguish wisdom from knowledge alone.

We will distinguish knowledge from judgment.

We will distinguish meaning from responsible participation.

We will distinguish coherence from moral goodness.

We will remember that a harmful system may preserve itself coherently.

We will declare every decision.

We will declare every decision-maker.

We will declare every affected person.

We will declare every absent stakeholder.

We will declare every future inheritor.

We will declare every alternative.

We will reveal which alternatives were never considered.

We will reveal which alternatives were not genuinely available.

We will distinguish formal choice from real capability.

We will declare every state of nature.

We will declare every probability.

We will declare every model.

We will declare every uncertainty.

We will distinguish risk from ambiguity.

We will distinguish ambiguity from deep uncertainty.

We will preserve unknown unknowns as a boundary of humility.

We will not let normalized probability pretend the model space is complete.

We will distinguish decision quality from outcome quality.

We will not allow luck to become proof of wisdom.

We will not allow misfortune to become proof of foolishness.

We will resist hindsight.

We will preserve the evidence available at the time.

We will declare every objective.

We will reveal who selected it.

We will reveal what it omits.

We will declare every value.

We will declare every weight.

We will declare every constraint.

We will preserve noncompensable boundaries.

We will not place dignity into a weighted sum as though enough convenience elsewhere could erase its violation.

We will distinguish expected loss from tail loss.

We will examine catastrophic consequence.

We will not permit an attractive average to hide who may be destroyed in the tail.

We will distinguish Bayesian judgment from worst-case judgment.

We will distinguish minimax from minimax regret.

We will distinguish robust action from morally legitimate action.

We will not call one decision criterion universal.

We will examine aspiration levels.

We will not allow satisficing to protect low ethical standards.'

We will recognize bounded rationality.

We will not demand impossible calculation from finite human beings.

We will improve the decision architecture rather than pretending limits do not exist.

We will examine the value of information.

We will seek information that can change action.

We will not collect information merely because collection is possible.

We will preserve consent.

We will preserve privacy.

We will examine the harm of delay.

We will examine the harm of premature action.

We will not demand certainty where reality cannot provide it.

We will not use uncertainty as permission to ignore credible danger.

We will use precaution proportionately.

We will distinguish precaution from fear.

We will distinguish precaution from prohibition.

We will examine the risks of action.

We will examine the risks of inaction.

We will examine reversibility.

We will measure option loss.

We will preserve future freedom where present evidence does not justify closure.

We will not make irreversible decisions merely because they are administratively convenient.

We will ask whether the aim is legitimate.

We will ask whether the action is effective.

We will ask whether the action is necessary.

We will search for less restrictive alternatives.

We will preserve proportionality.

We will preserve rights.

We will preserve appeal.

We will preserve explanation.

We will preserve remedy.

We will acknowledge moral uncertainty.

We will not force incomparable ethical frameworks onto one scale without justification.

We will seek actions defensible across serious moral perspectives.

We will not allow pluralism to become indifference.

We will preserve dignity as a constitutional floor.

We will preserve autonomy.

We will preserve informed consent.

We will protect those unable to exercise autonomy fully.

We will not confuse care with control.

We will not confuse autonomy with abandonment.

We will enlarge the boundary through compassion.

We will not erase healthy boundaries through compassion.

We will not decide for another person merely because we believe our intentions are good.

We will examine consequence.

We will permit harmed persons to describe harms absent from the model.

We will distinguish causal contribution from blame.

We will distinguish authority from control.

We will distinguish benefit from innocence.

We will identify who can repair.

We will not distribute responsibility until it disappears.

We will use contribution formulas carefully.

We will not convert Shapley allocation into moral verdict.

We will preserve accountability.

We will preserve decision records.

We will preserve assumptions.

We will preserve dissent.

We will preserve predictions.

We will compare prediction with outcome.

We will revise the model.

We will revise the institution.

We will provide remedy where harm occurred.

We will distinguish accountability from punishment alone.

We will join accountability to learning.

We will join learning to repair.

We will recognize stewardship.

We will remember that we received before we created.

We will distinguish stewardship from ownership.

We will distinguish preservation from freezing.

We will preserve fidelity while permitting form to mature.

We will protect Earth.

We will protect knowledge.

We will protect mathematics.

We will protect language.

We will protect cultural inheritance.

We will protect scientific integrity.

We will protect children.

We will not consume the whole inheritance merely because the present generation possesses the power to do so.

We will represent future generations.

We will preserve their freedom of choice.

We will not treat political absence as moral absence.

We will distinguish financial discounting from moral discounting.

We will not assign lesser dignity to a life because that life exists later.

We will preserve viable capability.

We will preserve natural resources.

We will preserve cultural heritage.

We will preserve institutional learning.

We will preserve peace.

We will preserve education.

We will examine the consequences of major projects before irreversibility is imposed.

We will govern commons through relationships.

We will declare access.

We will declare extraction.

We will declare renewal.

We will declare monitoring.

We will declare repair.

We will not privatize benefit while socializing loss.

We will not assume shared resources are incapable of organized governance.

We will preserve local knowledge.

We will preserve nested responsibility.

We will preserve adaptive rules.

We will recognize distributed expertise.

We will honor technical knowledge.

We will honor lived knowledge.

We will not let either erase the other.

We will examine correlated error.

We will not treat repeated agreement from one source as independent confirmation.

We will preserve principled dissent.

We will not romanticize dissent detached from evidence.

We will declare the time horizon.

We will declare the spatial scale.

We will examine local benefit.

We will examine global consequence.

We will examine present benefit.

We will examine future consequence.

We will not export loss beyond the chosen boundary.

We will preserve the Book I recognition:

Wisdom is coherent understanding faithfully lived.

We will preserve:

Knowledge explains. Meaning orients. Wisdom participates.

We will preserve the Principle of Living Coherence.

We will ask whether organized understanding strengthens life.

We will ask whether it preserves dignity.

We will ask whether it enlarges responsibility.

We will ask whether it protects future inheritance.

We will distinguish good intention from wise consequence.

We will join compassion to competence.

We will join competence to compassion.

We will preserve science.

We will preserve conscience.

We will preserve mathematics.

We will preserve moral boundaries mathematics cannot choose for us.

We will use artificial systems as instruments.

We will not hide human value decisions behind algorithms.

We will preserve human accountability.

We will preserve meaningful oversight.

We will preserve the right to explanation.

We will preserve the right to appeal.

We will preserve the right to correction.

We will distinguish computational optimization from lived wisdom.

We will preserve dual-aspect integrity.

We will examine exterior participation.

We will examine interior intention.

We will examine the boundary joining intention to consequence.

We will preserve the symbolic status of the Spirit Field.

We will not convert gratitude, hope, responsibility, forgiveness, wonder, purpose, compassion, or stewardship into an unsupported physical force.

We will permit symbolic language to remain meaningful because its status remains honest.

We will preserve gratitude without silencing correction.

We will preserve hope without denying evidence.

We will preserve responsibility without domination.

We will preserve forgiveness without erasing truth.

We will preserve wonder without requiring ignorance.

We will preserve purpose without fanaticism.

We will preserve compassion without paternalism.

We will preserve stewardship without ownership.

We will preserve civilization through teaching.

We will remember that books are not the only stones.

We will remember that equations are not the only stones.

We will remember that institutions are not the only stones.

Living persons remain the builders.

Every teacher may become another stone.

Every scientist may become another stone.

Every physician may become another stone.

Every engineer may become another stone.

Every parent may become another stone.

Every child who learns to love reality may become another stone.

We will not build monuments to ourselves.

We will leave foundations for persons we will never meet.

We will permit stronger evidence to correct us.

We will permit harmed persons to confront us.

We will permit future generations to judge what we leave.

We will remain teachable.

We will remain accountable.

We will remain compassionate.

We will remain stewards.

**Reality teaches.**

**Science discovers.**

**Mathematics describes.**

**The Architecture recognizes.**

**Knowledge explains.**

**Meaning orients.**

**Coherence preserves.**

**Judgment chooses.**

**Compassion enlarges responsibility.**

**Stewardship protects inheritance.**

**Accountability repairs.**

**Wisdom participates.**

**Civilization carries the Cathedral forward.**

## Beyond the Final Arch

The fourteen mathematical chambers now stand.

Reality entered mathematics through state, observation, measurement, model, and uncertainty.

Relation entered through functions, graphs, fields, networks, and dependence.

Differentiation entered through partition, symmetry breaking, classification, and recursive identity.

Boundary entered through topology, interface, flux, permeability, and controlled participation.

Vulnerability entered through perturbation, sensitivity, threshold, failure, and loss.

Adaptation entered through feedback, control, plasticity, reserve, and changing policy.

Attention entered through allocation, salience, competition, cost, and selection.

Awareness entered through integration, recurrence, belief state, global availability, temporal continuity, and report.

Learning entered through error, loss, update, gradient, plasticity, causality, and generalization.

Memory entered through retention, attractor, consolidation, retrieval, reconstruction, and inheritance.

Knowledge entered through representation, logic, evidence, probability, model, explanation, provenance, and correction.

Meaning entered through semantics, value, purpose, narrative, agency, relationship, and responsible orientation.

Coherence entered through generation, loss, viability, stability, resilience, reserve, debt, and the Cathedral Equation.

Wisdom has now entered through judgment, uncertainty, precaution, proportionality, irreversibility, accountability, compassion, stewardship, and future inheritance.

The mathematics has not completed reality.

It has disciplined the questions.

The equations have not produced wisdom.

They have revealed where wisdom remains necessary.

No further mathematical chamber can remove the final responsibility.

The learner must now become the participant.

## CATHEDRAL THREAD

Before wisdom could appear, reality had to become humanity's first teacher.

Before reality could become intelligible, relationships had to become visible.

Before relationships could produce identity, differentiation had to emerge.

Before identity could endure, boundaries had to regulate participation.

Before boundaries could matter, vulnerability had to reveal what could be lost.

Before vulnerable systems could continue, adaptation had to answer disturbance.

Before limited resources could be used intelligently, attention had to select.

Before selection could become lived presence, awareness had to organize experience.

Before experience could alter the future, learning had to change the learner.

Before learning could become history, memory had to preserve the change.

Before memory could become understanding, knowledge had to organize relationships.

Before knowledge could guide life, meaning had to identify what deserved devotion.

Before meaning could endure, coherence had to preserve indispensable relationships through change.

Before coherence could become civilization, wisdom had to judge what deserved preservation, what required transformation, who belonged within responsibility, and what must be passed forward.

The constitutional principle of the final chamber is now established:

**Wisdom is coherent judgment under uncertainty through which knowledge and meaning become compassionate, proportionate, accountable, and future-aware participation within reality.**

The practical-wisdom principle follows:

**Truthful reasoning must become joined to rightly ordered purpose, character, and action.**

The decision principle follows:

**A decision architecture must declare alternatives, evidence, models, uncertainty, values, constraints, affected persons, and future consequence before optimization can claim authority.**

The decision-quality principle follows:

**A good decision and a good outcome are related but distinguishable because luck may separate process from result.**

The bounded-rationality principle follows:

**Finite persons require finite search, aspiration levels, institutional support, and humility rather than fictional omniscience.**

The uncertainty principle follows:

**Expected loss, worst-case loss, regret, and robustness express different attitudes toward uncertainty; none becomes wisdom merely through mathematical consistency.**

The information principle follows:

**Additional information deserves pursuit when it can arrive in time, be obtained justly, and meaningfully improve the decision.**

The tail-risk principle follows:

**Average benefit must not conceal severe harm borne by comparatively few or by futures assigned low probability.**

The precautionary principle follows:

**Credible threats of serious or irreversible harm may enlarge responsibility even when full certainty remains unavailable.**

The reversibility principle follows:

**When evidence remains incomplete, preserving future options may possess value beyond immediate expected return.**

The proportionality principle follows:

**Effective means remain accountable to legitimate aims, necessity, burden, rights, and less restrictive alternatives.**

The moral-uncertainty principle follows:

**Uncertainty about ethical theory does not remove moral responsibility; it may support actions defensible across several serious moral perspectives.**

The dignity principle follows:

**Human standing must not be treated as one small compensable term inside an aggregate objective.**

The compassion principle follows:

**Another being's vulnerability becomes ethically visible when it enters the action architecture before harm occurs.**

The responsibility principle follows:

**Causal contribution, foreseeability, control, role, benefit, and repair are distinguishable dimensions of responsibility.**

The accountability principle follows:

**A wise decision leaves records, reasons, review pathways, correction, and remedy after authority has acted.**

The stewardship principle follows:

**Every generation receives before it creates and therefore becomes a temporary custodian rather than the absolute owner of inheritance.**

The intergenerational principle follows:

**Future persons possess moral relevance despite lacking present political power, and present action should preserve their basic capabilities and freedom of choice.**

The commons principle follows:

**Shared inheritance requires organized boundaries, monitoring, participation, learning, and repair rather than abandonment or extraction.**

The living-coherence principle follows:

**Coherence reaches its highest expression when organized understanding strengthens life, dignity, responsibility, and future inheritance.**

The dual-aspect principle follows:

**Interior integrity, exterior participation, and the truthful boundary joining intention to consequence may be represented distinctly without becoming separate realities.**

The Spirit Field principle follows:

**Interior wisdom may symbolically organize understanding, gratitude, hope, responsibility, forgiveness, wonder, purpose, compassion, and stewardship without being converted into an unearned physical mechanism.**

The civilization principle follows:

**Civilization is organized coherence transmitted through living persons who receive knowledge, preserve dignity, correct error, and prepare foundations for people they will never meet.**

The Book I principle now gathers its fullest life:

**Book I gave the Cathedral its constitutional soul.**

**Book II has given its chambers mathematical discipline.**

**Neither book is complete until understanding becomes participation.**

## THE FINAL INSCRIPTION

**Reality remains humanity's first teacher.**

**Science remains humanity's greatest discipline of discovery.**

**Mathematics remains humanity's most precise language of relationship.**

**Coherence remains the discipline through which indispensable organization endures.**

**Wisdom remains coherent understanding faithfully lived.**

**Compassion enlarges the boundary.**

**Stewardship protects the inheritance.**

**Civilization prepares the future.**

**The Cathedral belongs to Reality.**

**We remain its students.**

**We become its builders only through the faithfulness of our participation.**

# PART III

# THE DUAL-ASPECT FORMALISM

## From Coupled State Spaces to a Testable Research Architecture

# INTERLUDE

# FROM THE CATHEDRAL TO THE FORMAL LABORATORY

The fourteen chambers are complete.

The next task is not to add another chamber.

It is to formalize the relationships among them.

Reality has become state.

Relation has become structure.

Differentiation has become distinguishability.

Boundary has become regulated interface.

Vulnerability has become perturbation, sensitivity, threshold, and loss.

Adaptation has become changing policy.

Attention has become allocation.

Awareness has become modeled interior organization.

Learning has become update.

Memory has become retention.

Knowledge has become inference.

Meaning has become orientation.

Coherence has become generation and loss.

Wisdom has become responsible participation.

The Constitutional Cathedral has therefore reached completion.

Part III does not place another stage above Wisdom.

It turns toward the entire structure and asks:

Can the exterior and interior descriptions developed across the preceding chambers be joined within one formal architecture?

Can their variables be defined?

Can their observations be separated from their hidden states?

Can their couplings be estimated?

Can competing models be compared?

Can conditions of observability, identifiability, prediction, and failure be declared?

Can the formalism reveal where evidence ends and architectural interpretation begins?

Book I prepared this movement through Map VII—Interior and Exterior:

Exterior
Reality → Observation → Science → Mathematics → Architecture → Interior
Participation → Meaning → Wisdom → Compassion → Stewardship.

Its caption states:

Reality is encountered externally and lived internally. The Architecture recognizes the bridge between the measurable and the meaningful.

That bridge now enters formal mathematics.

Yet the constitutional order remains unchanged:

Empirical Foundation → Mathematical Language → Architectural Recognitio

The Architecture continues to inherit before it interprets. Its equations summarize or model relationships; they do not silently replace the empirical equations, methods, or evidentiary standards of the disciplines from which those relationships are inherited.

Part III therefore enters the Formal Laboratory under one oath:

No state will be confused with its measurement.

No latent variable will be confused with an independently discovered substance.

No correlation will be confused with identity.

No mathematical coupling will be treated as proof of metaphysical causation.

No first-person report will be dismissed merely because it is interior.

No exterior measurement will be permitted to claim more than its observation map supports.

No formalism will be permitted to become greater than the reality it attempts to describe.

# JUDGMENT, UNCERTAINTY, CONSEQUENCE, AND REPAIR



*Judgment acts without pretending uncertainty has disappeared.*

**JUDGMENT != CERTAINTY | OUTCOME != DECISION QUALITY**

FS-G20.01 — EQ-23.01 · EQ-23.02

### WISDOM CONSTITUTION AND ACTION BOUNDARY

Begin with coherent understanding faithfully lived; separate knowledge from responsible action.

FS-G20.02 — EQ-23.03 · EQ-23.07

### STAKEHOLDERS, OPTIONS, PERMISSIONS

Declare affected persons, feasible and accessible options, constraints, consent, dignity, and permission.

FS-G20.03 — EQ-23.03 · EQ-23.05

### FORECASTS, SCENARIOS, CALIBRATION

Preserve assumptions, probabilities, ranges, model uncertainty, calibration, and the value of further evidence.

FS-G20.04 — EQ-23.06 · EQ-23.08

### RISK AND DISTRIBUTED CONSEQUENCE

Keep severity, uncertainty, reversibility, unequal burden, externality, and affected-person vectors visible.

FS-G20.05 — EQ-23.07 · EQ-23.10

### PROTECTED GOODS AND FUTURE OPTIONS

Prevent dignity and indispensable goods from becoming ordinary prices; preserve future choice.

FS-G20.06 — EQ-23.04 · EQ-23.06

### ROBUST JUDGMENT AND REGRET

Compare expected value with minimax regret, tail risk, precaution, and irreversible-harm limits.

FS-G20.07 — EQ-23.05

### VALUE OF INFORMATION AND TIMING

Ask whether more evidence can change the decision before delay itself becomes harmful.

FS-G20.08 — EQ-23.03 · EQ-23.12

### ACTION, RESTRAINT, MONITORING

Record action, omission, restraint, execution, review conditions, stopping rules, and policy revision.

FS-G20.09 — EQ-23.12 · EQ-23.13

### DECISION VERSUS OUTCOME QUALITY

Separate responsible ex ante judgment from favorable or unfavorable realized outcomes.

FS-G20.10 — EQ-23.08 · EQ-23.11

### RESPONSIBILITY AND INHERITANCE

Audit knowledge, control, foreseeability, benefit, burden, authority, and intergenerational consequence.

FS-G20.11 — EQ-23.09 · EQ-23.12

### ACCOUNTABILITY, APPEAL, REPAIR

Preserve review, correction, restitution, restoration, reconciliation, and repair sufficiency.

FS-G20.12 — EQ-23.10 · EQ-23.11 · EQ-23.13 · EQ-23.14

### WISDOM AND CIVILIZATION PROFILE

Preserve future options, stewardship, quality profiles, civilizational participation, learning, and revision.

**DIGNITY | UNCERTAINTY | SOURCE TRANSPARENCY | FAILURE | REVISION**

FS identifies figure position; EQ identifies canonical equation. Reality remains the measure.

CHAPTER TWENTY-FOUR

# EXTERIOR AND INTERIOR STATE SPACES

*Coupled Domains, Interfaces, Observability, and the Mathematics of Two Descriptions*

*The First Formal Laboratory*

Reality is encountered through measurable relations.

Life is also lived through interior significance.

The formal problem is not to erase either description, but to determine how they may constrain one another faithfully.

## Guiding Question

How can exterior empirical states and interior lived states be represented within a mathematically disciplined formalism; how are hidden states related to observations, reports, actions, and boundaries; under what conditions are those states observable, identifiable, estimable, controllable, or fundamentally indistinguishable; and what prevents a dual-aspect model from becoming either an unsupported metaphysical dualism or a reduction that erases lived experience?

The fourteen mathematical chambers developed the required components separately.

Exterior states appeared through:

- physical variables;
- biological variables;
- neural activity;
- bodily regulation;
- behavior;
- environments;
- institutions;
- measurable consequences.

Interior organization appeared through:

- attention;
- awareness;
- affect;
- pain;
- memory;
- interpretation;
- intention;

- confidence;
- value;
- meaning;
- purpose;
- wisdom.

Until now, these variables were introduced where each constitutional chamber required them.

Part III must gather them.

The central movement is:

$$\text{One Participating Life} \rightarrow \{\text{ Exterior Empirical Description; Interior Lived Description }\} \rightarrow \text{Coupled Formal Architecture } [SDAE]$$

This does not begin by claiming that the two descriptions are:

- two physical universes;
- two substances;
- two spatial dimensions;
- two independent causal worlds;
- two complete and interchangeable accounts.

It begins with a more disciplined observation:

Human participation presents at least two epistemically distinguishable descriptions.

One concerns what can be publicly observed, instrumentally measured, behaviorally recorded, and causally manipulated.

The other concerns what is lived, intended, remembered, valued, suffered, hoped, and understood from within.

The formal challenge is to preserve the difference without converting difference into isolation.

## The Inheritance from Boundary

Book I established that boundaries are not merely separating edges. They are organized interfaces between interior coherence and exterior reality. They define what enters, what leaves, what becomes regulated, and how identity participates without dissolving.

Chapter Twenty-Four therefore begins where Chapter Four once stood:

$$\text{Interior} \leftrightarrow \text{Boundary} \leftrightarrow \text{Exterior}\,.$$

But Part III now asks more technical questions.

What is an interior state?

What is an exterior state?

Which of them is directly observed?

Which is inferred?

Which variables belong to the boundary?

Can one state be reconstructed from another?

Can distinct hidden states produce the same observations?

Can the same interior report arise through different exterior mechanisms?

Can the same exterior measurement accompany different lived states?

What evidence would distinguish a genuinely coupled dual-aspect model from an exterior-only model with a report readout?

These questions define the first Formal Laboratory.

## Terminological Discipline

| Term | Meaning in Part III |
|---|---|
| State | A mathematical representation containing the information required, under a declared model, to determine future evolution given future inputs and disturbances |
| State variable | One coordinate or component used to represent a system's state |
| State space | The set or manifold of states permitted by the model |
| Trajectory | An ordered sequence or curve of states through time |
| Exterior reality | Reality considered through publicly accessible empirical relations, measurements, behaviors, environments, and consequences |
| Exterior state | A modeled state composed of physical, biological, neural, behavioral, environmental, or institutional variables |
| Interior life | Lived experience considered through awareness, affect, meaning, intention, memory, value, and first-person participation |
| Interior state | A modeled representation of selected dimensions of lived experience; not a newly discovered physical substance or spatial dimension |
| Joint state | A mathematical state containing exterior, interior, and interface components within one formal model |
| Latent state | A state not directly available in the recorded observations and therefore inferred through a model |
| Observable | A measured or recorded quantity produced through an observation map |
| Observation map | A rule relating hidden state to measured output |
| Measurement | A value produced through a declared instrument, coding rule, or operational procedure |
| Report | An exterior communicative event used as evidence concerning an interior state |
| Behavior | Observable action that may depend upon exterior and interior organization |
| Interface state | A representation of bodily, sensorimotor, linguistic, communicative, and relational processes joining interior and exterior descriptions |
| Input | A declared influence entering the modeled dynamics |
| Control input | An input intentionally selected to influence the system |
| Disturbance | An influence not selected by the modeled controller |
| Noise | Unresolved variability assigned to state evolution, observation, or report |
| Parameter | A comparatively fixed quantity governing model structure or dynamics |
| Model class | A family of models sharing a declared mathematical architecture |

| Term | Meaning in Part III |
|---|---|
| Realization | A state-space system producing a specified input–output relation |
| Minimal realization | A realization using no unnecessary uncontrollable or unobservable state dimensions |
| Observability | Whether distinct internal states can be distinguished from available inputs and outputs over a declared interval |
| Estimability | Whether hidden states can be inferred with bounded or improving uncertainty under a declared model |
| Detectability | Whether unobservable state components that could prevent convergence are absent or sufficiently stable under a declared definition |
| Controllability | Whether admissible inputs can move the system among declared states within a specified model |
| Reachability | Whether a target state can be reached from a declared initial state |
| Identifiability | Whether parameters or model structures can be uniquely distinguished from available observations |
| Structural identifiability | Identifiability under ideal, noise-free observations assumed by the mathematical model |
| Practical identifiability | Ability to estimate quantities precisely from finite, noisy, realistically sampled data |
| Indistinguishable states | Distinct modeled states producing identical observable histories under the admissible experiments |
| State estimator | A mathematical process that reconstructs or predicts hidden states from observations and inputs |
| Observer | A dynamical system constructed to estimate the state of another system |
| Belief state | A probability distribution over hidden states conditional upon available history |
| First-person data | Disciplined reports or judgments supplied from the participant's experiential perspective |
| Third-person data | Measurements accessible to external observers through instruments, behavior, records, or physical observation |
| Correspondence | A modeled relation connecting exterior and interior states or observations |
| Coupling | A term through which the evolution of one modeled component depends upon another |
| Bidirectional coupling | A model in which exterior and interior variables influence each other's predicted evolution |
| Readout model | A model in which interior variables or reports depend upon exterior states without reciprocal influence |
| Common-cause model | A model in which exterior and interior observations are generated by another latent process |
| Projection | A map from a joint state to one of its exterior, interior, or interface descriptions |
| Compatibility relation | The set of exterior–interior state pairs permitted by a model |
| Dual-aspect formalism | A mathematical architecture preserving exterior and interior descriptions of one participating system without assuming that either description alone is complete |
| Reduction | An account claiming that one description is fully recoverable from or identical to another under declared conditions |

| Term | Meaning in Part III |
|---|---|
| Nonreduction | Refusal to claim complete equivalence where the evidence and mappings do not establish it |
| Coarse-graining | Representation that preserves selected large-scale relationships while omitting finer distinctions |
| Triangulation | Joint inference using several partially independent observation channels |
| Model comparison | Evaluation of competing formal architectures through prediction, evidence, intervention, simplicity, and failure conditions |
| Epistemic asymmetry | Difference between how a state is accessible to the experiencing participant and how it is accessible to an external observer |
| Dual-aspect coherence | Compatibility among exterior function, interior integrity, and the interface joining them |

These distinctions must remain visible.

$$\text{State} \neq \text{Observation} \quad \text{Exterior State} \neq \text{Everything Outside the Skin}$$

$$\text{Interior State} \neq \text{A Spatial Interior Dimension}$$

$$\text{Report} \neq \text{Experience Itself}$$

$$\text{Neural Correlate} \neq \text{Complete Lived Experience}$$

$$\text{Latent Variable} \neq \text{Independently Verified Entity}$$

$$\text{Observability} \neq \text{Transparency} \quad \text{Controllability} \neq \text{Ethical Permission}$$

$$\text{Model Coupling} \neq \text{Proven Metaphysical Causation}$$

$$\text{Two Descriptions} \neq \text{Two Universes}$$

## Scientific and Mathematical Foundations

This chapter stands upon the inheritance of:

- dynamical systems;
- control theory;
- state estimation;
- system identification;
- probability;
- Bayesian filtering;
- differential geometry;
- neuroscience;
- psychophysics;
- cognitive science;
- phenomenology;
- neurophenomenology;
- measurement theory.

Modern state-space methods became especially influential through the work of Rudolf Kalman. His filtering framework represented evolving hidden states separately from noisy observations and provided a recursive method for prediction and correction. His later system-theoretic work clarified controllability, observability, realization, and the fact that external input–output behavior determines only the portions of a linear system that are both controllable and observable.

This inheritance gives Part III its first mathematical discipline:

A system's state is not identical to what an instrument records.

The instrument records an output.

The state is inferred through a model relating output, input, dynamics, and uncertainty.

Nonlinear systems require additional care. Hermann and Krener developed differential-geometric observability conditions involving Lie derivatives and the local rank of an observation codistribution. These conditions show that observability is dependent upon the dynamics, the observation functions, and the region of state space being examined; it is not an unconditional property of a named variable.

Takens demonstrated that, under specific smooth, deterministic, generic conditions, the geometry of a finite-dimensional dynamical system may be reconstructed through delayed observations of a scalar measurement. This is a profound result, but it does not imply that any arbitrary time series reveals every hidden cause, nor that subjective meaning can be reconstructed automatically from one physiological signal.

Neuroscience also routinely distinguishes hidden states from measured signals. Dynamic causal modeling, for example, represents latent neuronal dynamics and a separate forward model through which those dynamics generate measured neuroimaging or electrophysiological data. The measured signal is therefore not simply declared identical to the underlying neuronal state.

The study of experience introduces another evidentiary dimension. Varela proposed neurophenomenology as a disciplined mutual constraint between first-person investigation and cognitive neuroscience rather than either an elimination of experience or an immunity of experience from scientific inquiry. Lutz and colleagues later used participants' structured first-person reports to group trials and examined corresponding differences in measured neural dynamics during a visual task. This did not transform reports into infallible observations, but it demonstrated that disciplined first-person information can alter the organization of empirical analysis.

Partially observable decision theory contributes another useful architecture. Smallwood and Sondik formalized control when the true internal state is not directly observed and decisions must instead be made from a probability distribution over possible states. The resulting belief state is a sufficient decision variable within the declared model, but it remains a probability representation used by the observer —not the hidden state itself and not necessarily a conscious belief.

These scientific inheritances establish the formal problem.

The exterior may be only partially observed.

The interior may be only partially reportable.

The boundary transforms both.

The model must remain explicit about what is known, what is inferred, and what remains inaccessible.

## Architectural Recognition

Only after empirical science and mathematics have spoken does the Architecture introduce the first recognition of Part III:

The Dual-Aspect Formalism represents one participating system through exterior empirical states, interior modeled states, and the boundary processes through which the two descriptions become mutually constraining.

Let:

$$X_E$$

be the exterior state space.

Let:

$$X_I$$

be the interior state space.

Let:

$$B$$

be the interface state space.

A first approximation is the product:

$$J = X_E \times X_I \times B.$$

A joint state is:

$$\zeta(t) = (x_E, (t), x_I(t) b(t)) \in J.$$

But the formalism need not assume that every exterior state can pair with every interior state.

A more disciplined joint space is:

$$J \subseteq X_E \times X_I \times B.$$

Only compatible triples belong to the modeled system.

Define projections:

$$\pi_E : J \to X_E, \pi_I : J \to X_I, \pi_B : J \to B.$$

Then:

$$x_E = \pi_E(\zeta), x_I = \pi_I(\zeta), b = \pi_B(\zeta).$$

### Equation Passport 24.1

*The Joint Dual-Aspect State Space*

$$J \subseteq X_E \times X_I \times B \zeta = (x_E, x_I, b)$$

Status: $[ARC / MOD]$

Exterior state: $x_E$

Interior modeled state: $x_I$

Interface state: $b$

Joint formal state: $\zeta$

Projection maps:

$$x_E = \pi_E(\zeta), x_I = \pi_I(\zeta), b = \pi_B(\zeta).$$

What it provides: One mathematical bookkeeping architecture for two descriptions and their interface

What it does not establish:

- two substances;
- a new physical dimension;
- a soul coordinate;
- one-to-one mapping;
- metaphysical identity;
- causal independence.

Architectural significance: Exterior and interior descriptions are preserved as distinguishable projections of one formal participation without assuming that either projection contains the whole.

## What Is a State?

A state is not merely a list of everything presently measurable.

A state is defined relative to a model.

For a deterministic discrete-time system:

$$x_{t+1} = F(x_t, u_t),$$

the present state $x_t$, together with the future input $u_t$, determines the next modeled state.

For a stochastic system:

$$p(x_{t+1}, \mid x_{0:t} u_{0:t}) = p(x_{t+1}, \mid x_t u_t).$$

This is a Markov-state condition.

The state summarizes the relevant modeled history.

It need not contain every historical detail.

It must contain enough of the past to support the declared prediction of the future.

Therefore:

State Sufficiency Is Model-Relative.

A state sufficient for predicting movement over one second may be insufficient for predicting biological adaptation over one decade.

A state sufficient for predicting a verbal response may be insufficient for explaining why that response matters to the speaker.

## State Spaces Are Not Always Euclidean

A simple state space may be:

$$X = R^n.$$

But not every state is naturally represented by an unrestricted vector.

A state space may be:

- a simplex;
- a graph;
- a manifold;
- a discrete category set;
- a constrained product;
- a probability-distribution space;
- a hybrid discrete–continuous system.

A probability state belongs to:

$$\Delta^{n-1} = \left\{ p, \in R^n : p_i \geq 0 \sum_i p_i = 1 \right\}.$$

An orientation may lie on a circle:

$$S^1.$$

A bodily posture may lie on a nonlinear manifold.

An interior-state representation involving affect, attention, intention, and self-model may require several different coordinate types.

The convenience of:

$$R^n$$

must not become an ontological claim that every meaningful distinction is linear, additive, or globally measurable.

## The Exterior State Space

The exterior state space may contain selected variables from:

- physics;
- physiology;
- neural activity;
- endocrine activity;
- immune function;
- behavior;
- movement;
- environment;
- social interaction;
- institutional structure.

Write:

$$x_E(t)=\begin{bmatrix} x_{phys}(t) \\ x_{neural}(t) \\ x_{body}(t) \\ x_{behavior}(t) \\ x_{environment}(t) \\ x_{social}(t) \end{bmatrix}.$$

This does not mean every study must include every block.

The state must be typed by the question.

For one experiment:

$$x_E$$

may contain only a small neuronal population and a stimulus condition.

For another:

$$x_E$$

may contain whole-body physiological regulation.

For another:

$$x_E$$

may contain a person, family, institution, and environment.

The word exterior does not mean merely external to the body.

Neural states and internal physiology belong to the exterior empirical description because they can, in principle, be approached through publicly shareable measurement and scientific inference.

## Exterior State Is Often Hidden

Even exterior physical states are rarely measured completely.

A blood sample reveals selected quantities.

An electroencephalogram records voltage differences at a limited set of locations.

A neuroimaging signal is transformed through physical and physiological processes before becoming data.

A behavioral record captures selected actions.

The true modeled exterior state may therefore be latent:

$$x_E \, not \equiv y_E.$$

Instead:

$$y_E=h_E(x_E,b,\theta_E)+\varepsilon_E.$$

Here:

- $y_E$ is the exterior observation;

- $h_E$ is the observation map;
- $b$ contains relevant interface processes;
- $\theta_E$ contains measurement parameters;
- $\varepsilon_E$ represents unresolved observation error.

## Equation Passport 24.2

***Exterior Observation***

$$y_E = h_E(x_E, b, \theta_E) + \varepsilon_E$$

Status: $[MOD]$

Exterior state: $x_E$

Measured output: $y_E$

Observation map: $h_E$

Interface contribution: $b$

Measurement parameters: $\theta_E$

Observation error: $\varepsilon_E$

Required declarations:

- instrument;
- sampling rate;
- spatial resolution;
- temporal resolution;
- calibration;
- preprocessing;
- missingness;
- noise assumptions.

What it prevents: Treating the recorded signal as an unmediated copy of reality
Architectural significance: Exterior reality becomes data only through an observation architecture.

# The Interior State Space

The interior state space represents selected dimensions of lived participation.

A provisional interior state may be written:

$$x_I(t) = \begin{bmatrix} a(t) \\ p(t) \\ e(t) \\ m(t) \\ v(t) \\ g(t) \\ s(t) \\ \mu(t) \end{bmatrix},$$

where:

- $a$ is attention;
- $p$ is perceptual content;
- $e$ is affect or felt condition;
- $m$ is presently active memory;
- $v$ is value orientation;
- $g$ is goal or intention;
- $s$ is self-model or identity organization;
- $\mu$ is experienced meaning.

This vector is illustrative.
It is not proposed as the universal coordinate system of human interiority.
One study may operationalize:

- pain intensity;
- pain unpleasantness;
- attention;
- expectation.

Another may operationalize:

- perceptual confidence;
- visual content;
- task intention.

Another may examine:

- autobiographical meaning;
- purpose;
- value conflict;

- hope.

The selected interior variables must always be declared.

## Interior Does Not Mean Spatially Inside

The phrase interior state space is easily misunderstood.

It does not refer merely to variables located inside the skull or body.

Neural tissue is physically interior to the body but belongs to the exterior empirical description.

A remembered childhood event may concern an exterior historical situation but is presently lived through the interior description.

The interior–exterior distinction is therefore not simply:

$$\text{Inside Skin versus Outside Skin}.$$

It is primarily a distinction between modes of access and description:

$$\text{Publicly Measurable Description and Lived Participatory Description}.$$

## The Interior State Space Is Modeled

No third-person observer directly opens an instrument and reads:

$$x_I.$$

Interior variables are approached through:

- first-person report;
- structured introspection;
- forced-choice judgment;
- confidence rating;
- narrative;
- behavior;
- memory;
- physiological relation;
- context;
- intervention.

The model may infer:

$$\hat{x}_I.$$

But:

$$\hat{x}_I \neq x_I$$

unless the model has earned that equivalence under declared conditions.

The estimate carries uncertainty.

The experience remains greater than the estimate.

## The Experiencing Person Is Not Omniscient About the Interior

First-person access has a distinctive authority.

A person is ordinarily the primary witness to whether pain, fear, grief, or a visual experience is present.

Yet first-person access need not be:

- complete;
- perfectly verbalizable;
- causally explanatory;
- historically infallible;
- immune from expectation;
- immune from social influence.

A person may know:

I am in pain.

The person may not know:

- the complete physiological cause;
- the long-term prognosis;
- the population-level mechanism;
- whether two processes are contributing;
- how language has shaped the report.

The formalism therefore protects both:

$$\text{First-Person Authority}$$

and:

$$\text{First-Person Fallibility}.$$

## The Interface State

Book I's boundary was an organized interface.

Part III now gives that interface a state:

$$b(t) \in B.$$

The interface may include:

- sensory transduction;
- interoception;
- motor action;
- autonomic response;
- facial expression;
- vocal production;

- language;
- gesture;
- attention to internal events;
- social context;
- measurement instruments;
- reporting rules.

The interface is not a third metaphysical substance.
It is the formal region through which the two descriptions become related.
A stimulus becomes neural and bodily activity through an interface.
An intention becomes action through an interface.
An experience becomes report through an interface.
A report becomes another person's evidence through an interface.

The Boundary Does Not Merely Separate Descriptions.
It Organizes Their Participation.

## The General Coupled Dynamics

Let:

$$u_E(t)$$

represent exterior intervention or control.
Let:

$$u_I(t)$$

represent a modeled interior regulatory input such as deliberately sustained attention, reappraisal, rehearsal, or selected intention.
Let:

$$d_E(t), d_I(t)$$

represent disturbances.
The coupled architecture is:

$$\dot{x}_E = f_E(x_E, x_I, b, u_E, d_E, t, \theta_E), \dot{x}_I = f_I(x_I, x_E, b, u_I, d_I, t, \theta_I),$$
$$\dot{b} = f_B(b, x_E, x_I, u_E, u_I, c, t, \theta_B).$$

Observations are:

$$y_E = h_E(x_E, b, \phi_E) + \varepsilon_E, r = h_R(x_I, x_E, b, \lambda, c, \phi_R) + \varepsilon_R,$$
$$y_A = h_A(x_E, x_I, b, c, \phi_A) + \varepsilon_A,$$

where:

- $r$ is report;
- $\lambda$ is language or response code;
- $c$ is context;

- $y_A$ is observable action.

## Equation Passport 24.3

*General Dual-Aspect Dynamics*

$$\dot{x}_E = f_E(x_E, x_I, b, u_E, d_E, t) \dot{x}_I = f_I(x_I, x_E, b, u_I, d_I, t)$$

$$\dot{b} = f_B(b, x_E, x_I, u_E, u_I, c, t)$$

Status: $[ARC/MOD/CONJ]$

Exterior state: $x_E$

Interior modeled state: $x_I$

Interface state: $b$

Exterior input: $u_E$

Interior regulatory input: $u_I$

Context: $c$

Required declarations:

- operational meaning of every state;
- whether coupling is causal or predictive;
- intervention availability;
- parameter units;
- delay;
- noise;
- model comparison;
- ethical permission.

What bidirectional terms do not establish: An immaterial mind pushing physical matter

Architectural significance: The formalism permits reciprocal dependence to be modeled while requiring every coupling to earn empirical interpretation.

# Interior Regulation Must Be Operationalized

The term:

$$u_I$$

must not become a hidden supernatural input.

It may represent experimentally operationalized conditions such as:

- an instruction to attend;
- learned meditation practice;
- cognitive reappraisal;

- deliberate recall;
- motor imagery;
- value-based choice;
- intention formation.

These processes are embodied.
The instruction arrives through exterior channels.
The learned capacity depends upon history.
The resulting interior regulation may alter behavior and measured physiology.
A model may assign the effect to an interior variable.
But that assignment must compete against models involving:

- exterior neural state alone;
- common latent causes;
- reporting expectation;
- unmeasured context.

# A Linear Coupled Prototype

Let:

$$z = \begin{bmatrix} x_E \\ x_I \\ b \end{bmatrix}.$$

A linear time-invariant prototype is:

$$\dot{z} = Az + Bu + w, y = Cz + v.$$

Partition:

$$A = \begin{bmatrix} A_{EE} & A_{EI} & A_{EB} \\ A_{IE} & A_{II} & A_{IB} \\ A_{BE} & A_{BI} & A_{BB} \end{bmatrix}.$$

Then:

$$\begin{aligned} \dot{x}_E &= A_{EE} x_E + A_{EI} x_I + A_{EB} b + B_E u + w_E, \\ \dot{x}_I &= A_{IE} x_E + A_{II} x_I + A_{IB} b + B_I u + w_I, \\ \dot{b} &= A_{BE} x_E + A_{BI} x_I + A_{BB} b + B_B u + w_B. \end{aligned}$$

## Equation Passport 24.4

***Linear Dual-Aspect Prototype***

$$\dot{z} = Az + Bu + w\, y = Cz + v$$

Status: $[MOD]$

Exterior self-dynamics: $A_{EE}$

Interior self-dynamics: $A_{II}$

Interface self-dynamics: $A_{BB}$

Exterior-to-interior coupling: $A_{IE}$

Interior-to-exterior coupling: $A_{EI}$

Boundary couplings: $A_{EB}, A_{IB}, A_{BE}, A_{BI}$

What the block structure permits: Direct comparison of alternative coupling hypotheses
What it does not guarantee:

- linearity;
- stationarity;
- correct dimensionality;
- causal interpretation;
- unique latent coordinates.

Architectural significance: The block matrix makes hidden assumptions about direction and coupling visible.

## Four Competing Model Classes

The Dual-Aspect Formalism should never begin by assuming that full bidirectional coupling is already established.
At least four model classes should be compared.

### Model $M_0$: Exterior-Only

$$\dot{x}_E = f_E(x_E, u, d) r = h_R(x_E, b, c) + \varepsilon_R.$$

Interior variables add no predictive information beyond exterior state and context.

### Model $M_1$: Interior Readout

$$x_I = g(x_E, h_t) \dot{x}_E = f_E(x_E, u, d) r = h_R(x_I, b, c) + \varepsilon_R.$$

Interior state summarizes exterior history but does not feed back into exterior dynamics.

### Model $M_2$: Coupled Dual-Aspect

$$\dot{x}_E = f_E(x_E, x_I, b, u, d) \dot{x}_I = f_I(x_I, x_E, b, u, d).$$

Interior and exterior modeled states each improve prediction of the other.

### Model $M_3$: Common Latent Process

$$\dot{z} = f_z(z, u, d) x_E = \pi_E(z), x_I = \pi_I(z).$$

Apparent cross-coupling is generated by another latent process rather than direct reciprocal dependence.

The scientifically appropriate question is not:

Which model best expresses our preferred philosophy?

It is:

Which model best predicts, explains, survives intervention, preserves parsimony, and reveals its own failure conditions?

## Equation Passport 24.5

*Competing Dual-Aspect Models*

$$M_0 : x_E \to r M_1 : x_E \to x_I \to r M_2 : x_E \leftrightarrow x_I M_3 : z \to \begin{cases} x_E \\ x_I \end{cases}$$

Status: $[MOD/CONJ]$

Comparison criteria may include:

- out-of-sample prediction;
- intervention response;
- parameter identifiability;
- model evidence;
- complexity;
- robustness;
- reproducibility.

Failure condition: Only the favored model is fitted

Architectural significance: Dual-aspect language becomes scientific only when exterior-only and alternative latent explanations remain genuine competitors.

# Observability

A hidden state is observable when distinct states produce distinguishable output histories under the admissible inputs.

Consider:

$$\dot{z} = Az, y = Cz.$$

The linear observability matrix is:

$$O = \begin{bmatrix} C \\ CA \\ CA^2 \\ \vdots \\ CA^{n-1} \end{bmatrix}.$$

The pair $(A, C)$ is completely observable when:

$$rank\left(O\right)=n.$$

The finite-horizon observability Gramian is:

$$W_O\left(T\right)=\int_0^T e^{A^T t} C^T C e^{At}\,dt.$$

Under the standard linear conditions, complete observability over the interval is associated with:

$$W_O\left(T\right)>0.$$

Kalman's realization framework made observability central to determining which internal state dimensions can be reconstructed from input–output behavior.

## Equation Passport 24.6

***Linear Observability***

$$O=\begin{bmatrix} C \\ CA \\ \vdots \\ CA^{n-1} \end{bmatrix} rank\left(O\right)=n.$$

Status: $\left[THM/MOD\right]$

State dynamics: $A$

Observation map: $C$

State dimension: $n$

Complete observability: Full rank

What observability depends upon:

- model;
- sensors;
- sampling;
- inputs;
- time horizon;
- noise;
- operating point.

What it does not establish:

- direct measurement;
- consciousness;
- causal importance;
- ethical permission to infer;
- uniqueness outside the model class.

Architectural significance: A hidden state becomes scientifically addressable only through the relationships by which it changes observable history.

## Observability Is Not an Intrinsic Label

A variable is not simply:

observable

or:

unobservable

for all possible purposes.

A state may be unobservable from:

- one sensor;
- one time point;
- one task;
- one passive recording.

It may become observable after:

- another sensor is added;
- a perturbation is introduced;
- the observation interval is extended;
- a different variable is recorded.

Therefore:

Observability Belongs to the System–Experiment Pair.

Not to the hidden variable alone.

## Interior Observability

Suppose the joint state is:

$$z = \begin{bmatrix} x_E \\ x_I \end{bmatrix}$$

and observations are:

$$y = \begin{bmatrix} y_E \\ r \\ y_A \end{bmatrix}.$$

Interior-state components are observable only when changes in $x_I$ create distinguishable changes in the available history of:

- report;
- action;

- exterior measurement;
- context-sensitive response.

If two interior states produce the same output history under every admissible experiment, the model cannot distinguish them.
They belong to one observational equivalence class.

# Indistinguishable States

For initial states $z_1$ and $z_2$, define:

$$z_1 \sim z_2$$

when:

$$y(t; z_1; u) = y(t; z_2; u)$$

for every admissible input $u$ over the declared interval.

Then:

$$[z] = \{z' : z' \sim z\}$$

is the equivalence class of states observationally indistinguishable from $z$.

The evidence may therefore support:

$$[z]$$

rather than one exact hidden state.
This is a major constitutional protection.
A model should not report:

The person is in interior state $x_I^{¿}$.

when the data justify only:

The observations are compatible with a family of interior states represented by $[x_I]$.

# Observability and Privacy

Interior partial observability is not merely a defect awaiting total elimination.
Some of it may express:

- limited language;
- incomplete self-access;
- scientific limitation;
- ethical privacy;
- contextual uniqueness.

A person should not be treated as fully transparent merely because a model produces a high-confidence estimate.

$$\text{Inferable} \Rightarrow \not\!\!\text{Owned by the Inferer}.$$

## Controllability

For the linear system:

$$\dot{z} = Az + Bu,$$

the controllability matrix is:

$$C = [B \quad AB \quad A^2B \quad \cdots \quad A^{n-1}B].$$

Complete controllability requires:

$$rank\ (C) = n.$$

A controllable state direction can, in principle, be influenced through the declared input.
An uncontrollable direction cannot be reached through that input under the model.

### Equation Passport 24.7

*Linear Controllability*

$$C = [B \quad AB \quad \cdots \quad A^{n-1}B] rank\ (C) = n.$$

Status: $[THM / MOD]$

Dynamics: $A$

Input map: $B$

State dimension: $n$

Complete controllability: Full rank
What it provides: A mathematical test of modeled state reachability
What it does not provide:

- ethical authority;
- informed consent;
- safety;
- clinical benefit;
- legitimate purpose.

Architectural significance: The ability to alter a state creates responsibility; it does not create permission.

## Controllability Is Not Desirability

A system may be controllable toward:

- healing;
- learning;
- fear;
- compliance;
- dependence;

- manipulation.

The mathematical property is value-neutral.
The ethical architecture is not.
Part III must therefore distinguish:

$$\text{Can Alter}$$

from:

$$\text{May Alter}$$

and:

$$\text{Should Alter}.$$

## State Estimation

Suppose:

$$z_{t+1} = A z_t + B u_t + w_t, y_t = C z_t + v_t,$$

with:

$$w_t \sim N(0, Q), v_t \sim N(0, R).$$

The Kalman prediction is:

$$\hat{z}_{t|t-1} = A \hat{z}_{t-1|t-1} + B u_{t-1}, P_{t|t-1} = A P_{t-1|t-1} A^T + Q.$$

The innovation is:

$$\nu_t = y_t - C \hat{z}_{t|t-1}.$$

Its covariance is:

$$S_t = C P_{t|t-1} C^T + R.$$

The gain is:

$$K_t = P_{t|t-1} C^T S_t^{-1}.$$

The updated estimate is:

$$\hat{z}_{t|t} = \hat{z}_{t|t-1} + K_t \nu_t.$$

The updated covariance is:

$$P_{t|t} = (I - K_t C) P_{t|t-1}.$$

Kalman's original framework gave a recursive solution to linear filtering and prediction under its stated stochastic assumptions.

## Equation Passport 24.8

***Recursive State Estimation***

$$\text{Prediction} \rightarrow \text{Innovation} \rightarrow \text{Correction}$$

$$\text{Status: } [THM / MOD]$$

Predicted state: $\hat{z}_{t \mid t-1}$

Updated state: $\hat{z}_{t \mid t}$

State uncertainty: $P_t$

Process noise: $Q$

Observation noise: $R$

Innovation: $\nu_t$

Required assumptions for the classical optimal result:

- linear dynamics;
- linear observation;
- Gaussian noise;
- correctly specified covariances.

Failure conditions:

- model-form error;
- nonlinear dynamics;
- non-Gaussian tails;
- biased measurements;
- changing parameters;
- invalid report model.

Architectural significance: Evidence should correct expectation in proportion to the estimated reliability of both.

## Precision Can Be Wrong

A filter can produce:

$$P_t \to 0$$

while its model is wrong.

Internal uncertainty becomes small.

External error remains.

This is overconfident estimation.

Therefore:

$$\text{Narrow Posterior} \neq \text{Faithful State Estimate}.$$

Model discrepancy must remain visible.

## The Belief State

When the true state is only partially observable, define:

$$\beta_t(z) = p(z_t, = z \mid y_{0:t} u_{0:t-1}).$$

Prediction gives:

$$\beta_{t|t-1}(z') = \int p(z', | z u_{t-1}) \beta_{t-1}(z) \, dz .$$

Observation update gives:

$$\beta_t(z') = \frac{p(y_t | z') \beta_{t|t-1}(z')}{\int p(y_t | \tilde{z}) \beta_{t|t-1}(\tilde{z}) \, d\tilde{z}} .$$

A policy may depend upon the belief state:

$$u_t = \pi(\beta_t).$$

In a partially observable Markov decision process, this probability distribution can serve as the sufficient state for planning within the model.

## Equation Passport 24.9

***Belief-State Formalism***

$$\beta_t(z) = p(z_t = z \mid \text{history})$$

Status: $[MOD]$

Hidden state: $z_t$

Belief state: $\beta_t$

Observation history: $y_{0:t}$

Action history: $u_{0:t-1}$

What it provides: A state of uncertainty for an observer or controller
What it does not establish:

- conscious belief;
- lived awareness;
- truth of the model;
- transparency of another person;
- moral authority to act.

Architectural significance: Under partial observability, the responsible state of the decision-maker is not certainty but an organized distribution of possibilities.

# The Belief State Is Not the Interior State

The observer may possess:

$$\beta_t(x_I).$$

The person possesses or lives through:

$$x_I(t).$$

These are not the same.

The first is the observer's uncertainty.

The second is the modeled interior condition.

Confusing them would place the model's belief in the position of the person's experience.

## Nonlinear Observability

Consider:

$$\dot{z}=f(z), y=h(z).$$

Define the first Lie derivative:

$$L_f h(z)=\nabla h(z)^T f(z).$$

Higher derivatives are:

$$L_f^2 h=L_f(L_f h),$$

and so forth.

Construct the local observation differential family:

$$dh, d L_f h, d L_f^2 h, \ldots$$

A prototype observability rank condition requires these differentials to span the cotangent space locally:

$$rank \begin{bmatrix} dh \\ d L_f h \\ \vdots \end{bmatrix} = n.$$

Hermann and Krener developed this differential-geometric approach for nonlinear systems. Its conclusions are local and dependent upon technical conditions; passing a rank test at one state does not establish global distinguishability across the whole state space.

### Equation Passport 24.10

*Nonlinear Observability Rank*

$$L_f h=\nabla h^T f rank(d, h, d, L_f hd L_f^2 h \ldots)=n.$$

Status: $[THM/MOD]$

Dynamics: $f$

Observation: $h$

State dimension: $n$

What it may support: Local distinguishability

What it does not guarantee:

- global observability;
- robustness to noise;

- practical estimation;
- valid interior-state interpretation;
- unique parameter recovery.

Architectural significance: Nonlinear state information may become visible through how observations change through time, not merely through one instantaneous measurement.

## Time Reveals What One Measurement Conceals

A single observation:

$$y_t$$

may be ambiguous.

A history:

$$y_t, y_{t-\tau}, y_{t-2\tau}, \dots$$

may contain enough relational structure to distinguish states.

Define the delay vector:

$$\Phi_{h,\tau}^{(k)}(z)=(h,(z),h,(\varphi_\tau z),h(\varphi_{2\tau} z)\dots h(\varphi_{k\tau} z)),$$

where:

$$\varphi_t$$

is the system flow.

Under the conditions of Takens's theorem, a generic smooth scalar observation of a compact $m$-dimensional deterministic system can yield an embedding in a delay space of dimension $2m+1$.

### Equation Passport 24.11

***Delay-Coordinate Reconstruction***

$$\Phi_{h,\tau}^{(2m)}(z)=(h,(z),h,(\varphi_\tau z)\dots h(\varphi_{2m\tau} z))$$

Status: $[THM/MOD]$

State manifold dimension: $m$

Observation function: $h$

Delay: $\tau$

Embedding dimension in the classical theorem: $2m+1$

Required conditions include:

- smooth deterministic dynamics;
- compact finite-dimensional manifold;
- generic observation;
- suitable sampling;

- adequate data.

What it does not establish:

- causal mechanism;
- exact state variables;
- applicability to arbitrary noisy systems;
- reconstruction of semantic meaning;
- reconstruction of consciousness from one signal.

Architectural significance: Historical observation may reveal state-space geometry that one isolated measurement cannot.

## The Takens Overreach Boundary

It would be incorrect to say:

A heart-rate time series contains the whole person.

Or:

An EEG channel reconstructs consciousness.

Or:

A social metric reconstructs civilization.

Delay reconstruction concerns the geometry of a declared dynamical system under specific mathematical assumptions.

It does not establish that every meaning-bearing aspect of the person belongs to that measured attractor.

## Realization and Coordinate Freedom

Suppose:

$$\dot{z} = Az + Bu, y = Cz.$$

Let:

$$\tilde{z} = Tz$$

for invertible $T$.

Then:

$$\dot{\tilde{z}} = TAT^{-1}\tilde{z} + TBu, y = CT^{-1}\tilde{z}.$$

Define:

$$\tilde{A} = TAT^{-1}, \tilde{B} = TB, \tilde{C} = CT^{-1}.$$

The transformed system generates the same input–output behavior.

Minimal linear realizations of the same input–output map are related through such coordinate transformations under the standard conditions.

### Equation Passport 24.12

***State-Coordinate Equivalence***

$$\tilde{z} = Tz \tilde{A} = TA T^{-1}, \tilde{B} = TB, \tilde{C} = C T^{-1}.$$

Status: $[THM / MOD]$

Transformation: $T$

Invariant: Input–output relation

What changes: Internal coordinates

What it reveals: Hidden-state coordinates are not automatically privileged merely because one fitted model uses them

Failure condition: A convenient latent axis is treated as a uniquely real psychological or biological entity

Architectural significance: The model may identify a relational structure more strongly than it identifies one coordinate language.

## Coordinate Humility

Suppose a latent model produces dimensions labeled:

- coherence;
- threat;
- meaning;
- self-control.

An invertible transformation may produce another coordinate system with equal predictive power.

The labels do not become natural kinds merely because the software returns four axes.

A latent coordinate earns interpretation through:

- intervention;
- convergent evidence;
- invariance;
- replication;
- relation to established constructs;
- explanatory necessity.

## Identifiability

Observability concerns hidden states.

Identifiability concerns parameters or model structures.

Let:

$$p_\theta(y_{0:T} \mid u_{0:T})$$

be the observable distribution generated by parameter $\theta$.

Structural identifiability requires:

$$p_\theta(y_{0:T} \mid u_{0:T}) = p_{\theta'}(y_{0:T} \mid u_{0:T})$$

to imply:

$$\theta = \theta'.$$

If:

$$\theta \neq \theta'$$

but both generate the same observations, the parameters are not identifiable from the experiment.

A coupling coefficient may appear in the equation and remain unidentifiable in practice.

Therefore:

$$\text{Parameter Exists in Model} \not\Rightarrow \text{Parameter Can Be Estimated}.$$

## Observability and Identifiability Must Not Be Confused

A state may be observable while a parameter is not identifiable.

A parameter may be identifiable from many trials while one instantaneous state remains uncertain.

A report model may be identifiable only after language, context, and response bias are included.

The formalism must state which problem is being solved.

## The Report Channel

Let:

$$x_I$$

be the interior state.

Let:

$$q$$

be the question or reporting prompt.

Let:

$$\lambda$$

be the language and response vocabulary.

Let:

$$c$$

be context.

Let:

$$m$$

be memory.

Let:

$$x_E$$

include the bodily and neural conditions supporting report production.

A report model is:

$$r = h_R(x_I, x_E, b, q, \lambda, c, m) + \varepsilon_R.$$

The report depends upon more than the experience.

A person may experience something but lack the vocabulary to report it.

The question may alter attention.

Memory may reconstruct.

Social context may encourage concealment.

Motor impairment may prevent expression.

The report is therefore evidence.

It is not a transparent duplicate.

## Equation Passport 24.13

***The Interior-to-Report Channel***

$$r = h_R(x_I, x_E, b, q, \lambda, c, m) + \varepsilon_R$$

Status: $[MOD]$

Interior state: $x_I$

Exterior enabling state: $x_E$

Boundary: $b$

Question: $q$

Language: $\lambda$

Context: $c$

Memory: $m$

Report error: $\varepsilon_R$

Possible report distortions:

- ambiguity;
- response bias;
- demand characteristics;
- memory error;
- fear;
- social pressure;
- motor limitation;
- scale misuse.

What the equation preserves: Experience and report remain related but distinguishable

Architectural significance: The path from living through an experience to communicating it is itself an organized boundary process.

## Silence Is Not Proof of Absence

If:

$$r=0,$$

it does not follow that:

$$x_I=0.$$

The person may:

- lack language;
- lack access;
- lack safety;
- lack motor control;
- choose privacy;
- misunderstand the question.

Likewise, a positive report does not by itself determine the complete mechanism.

Both errors must be resisted.

## Disciplined First-Person Data

Varela's neurophenomenological proposal treated first-person investigation as a disciplined methodological partner rather than an unstructured stream of private assertion. Lutz and colleagues operationalized a related strategy by using participants' reported cognitive context to organize trial classes and then testing whether measured neural dynamics differed among them.

A disciplined first-person protocol may therefore include:

- clear prompts;
- training in distinctions;
- temporal proximity to experience;
- repeated sampling;
- confidence reporting;
- vocabulary calibration;
- contradiction testing;
- independent exterior measurement.

The goal is not to transform experience into ordinary third-person data.

It is to improve the quality of the boundary through which experience becomes communicable evidence.

# Triangulation

Let observed evidence be:

$$D_t = \left(y_E, r, y_A, c\right)_t.$$

Joint inference is:

$$p\left(x_E, x_I, b \mid D_{0:t}\right) \propto p\left(D_t, \mid, x_E x_I b\right) p\left(x_E, x_I, b \mid D_{0:t-1}\right).$$

Under a justified conditional-independence approximation:

$$p\left(D_t, \mid, x_E x_I b\right) = p\left(y_E, \mid x_E b\right) p\left(r, \mid, x_I, x_E bc\right) p\left(y_A, \mid, x_I, x_E bc\right).$$

## Equation Passport 24.14

***Multimodal Triangulation***

$$p\left(x_E, x_I, b, \mid, y_E, r\, y_A c\right)$$

Status: $\left[MOD\right]$

Exterior measurement: $y_E$

Report: $r$

Action: $y_A$

Context: $c$

Hidden states: $x_E, x_I, b$

Possible benefit: Each evidence channel constrains different ambiguities
Required caution: Conditional independence must not be assumed merely for computational convenience
Failure conditions:

- all channels share one bias;
- report influences behavior used as independent evidence;
- preprocessing leaks labels;
- context is omitted;
- the same source is counted repeatedly.

Architectural significance: No single observation channel is granted automatic supremacy; credibility emerges through organized convergence and visible disagreement.

# Disagreement Among Channels Is Information

Suppose:

$$r$$

indicates severe pain.

Suppose one exterior test shows no detectable tissue injury.

The incoherent conclusion is:

No injury was measured; therefore no pain exists.

Another incoherent conclusion is:

Pain was reported; therefore the proposed physical cause is established.

The disciplined conclusion is:

- the pain report is evidence of lived pain;
- the exterior test constrains one class of physical explanations;
- other mechanisms may remain;
- further investigation may be required.

The disagreement reveals that:

$$\text{Experience}$$

and:

$$\text{Proposed Cause}$$

were being confused.

## Dynamic Causal Modeling as an Exterior Example

Dynamic causal modeling uses hidden neuronal states:

$$z_N$$

with dynamics:

$$\dot{z}_N = f_N(z_N, u, \theta_N),$$

and a measurement model:

$$y_{scan} = h_{hemo}(z_N, \theta_H) + \varepsilon .$$

The measured brain-imaging signal is generated through a forward model rather than equated directly with neuronal activity.

This offers Part III a methodological lesson.

Even within the exterior empirical domain:

$$\text{Measured Signal} \neq \text{Hidden Mechanism} .$$

Therefore the inference from measured signal to lived interior state requires at least one additional relationship, not fewer.

## Two Descriptions of One Formal Process

The Dual-Aspect Formalism may be written through one joint process:

$$\zeta(t) \in J .$$

Exterior and interior descriptions are projections:

$$x_E(t) = \pi_E(\zeta(t)), x_I(t) = \pi_I(\zeta(t)).$$

This says:

- the descriptions belong to one formal history;
- each preserves different relationships;
- neither projection is automatically complete.

It does not say what $\zeta$ ultimately is metaphysically.

The joint space is a mathematical instrument.
It may later be interpreted through:

- an exterior-only ontology;
- a nonreductive physicalism;
- neutral monism;
- another philosophical framework.

Part III need not settle that dispute before defining what evidence would distinguish models.

## Equation Passport 24.15

*Two Projections, One Formal History*

$$\zeta \in J x_E = \pi_E(\zeta) x_I = \pi_I(\zeta)$$

Status: $[ARC/MOD/CONJ]$

Joint formal process: $\zeta$

Exterior projection: $\pi_E$

Interior projection: $\pi_I$

What it preserves:

- one participating history;
- descriptive distinction;
- possible non-equivalence;
- empirical openness.

What it leaves unresolved:

- ultimate ontology;
- complete reduction;
- fundamental causal direction;
- whether one projection can eventually be derived from the other.

Architectural significance: Formal unity need not require descriptive erasure.

## Fibered Compatibility

For an exterior state $x_E$, define the set of compatible interior states:

$$F_I(x_E) = \{x_I : (x_E, x_I) \in R_{EI}\},$$

where:

$$R_{EI} \subseteq X_E \times X_I$$

is the compatibility relation.
The set:

$$F_I(x_E)$$

is the interior fiber over $x_E$.

If:

$$|\ F_I(x_E)\ | > 1,$$

the exterior state does not uniquely determine the interior state at the model's resolution.
Similarly:

$$F_E(x_I) = \{x_E : (x_E, x_I) \in R_{EI}\}.$$

Several exterior mechanisms may be compatible with one reported interior condition.

### Equation Passport 24.16

***Exterior–Interior Compatibility Fibers***

$$F_I(x_E) = \{x_I : (x_E, x_I) \in R_{EI}\} F_E(x_I) = \{x_E : (x_E, x_I) \in R_{EI}\}$$

Status: $[ARC/MOD]$

Compatibility relation: $R_{EI}$

Exterior fiber: $F_E(x_I)$

Interior fiber: $F_I(x_E)$

One-to-one mapping condition:

$$|\ F_I(x_E)\ | = 1$$

and:

$$|\ F_E(x_I)\ | = 1$$

for all relevant states.
Why this matters: One-to-one equivalence must be demonstrated, not assumed
Architectural significance: The formalism can represent lawful correspondence without forcing identity.

## Coarse Measurement Enlarges the Fiber

Suppose the true exterior state is high-dimensional:

$$x_E^{¿}.$$

The instrument records:

$$y_E = Q(x_E^{¿}),$$

where $Q$is a coarse-graining map.

Many states may satisfy:

$$Q(x_E) = y_E.$$

Therefore one measured exterior output may correspond to a large set of exterior states and an even larger set of compatible interior states.
Lack of one-to-one mapping may arise from:

- true biological multiplicity;
- measurement limitation;
- omitted context;
- person-specific history;
- model misspecification.

The formalism should not attribute every ambiguity to metaphysical mystery when ordinary measurement limitation is sufficient.

## Cross-Domain Compatibility

Define a compatibility score:

$$\kappa(x_E, x_I, b, c) \geq 0.$$

Lower values indicate stronger compatibility under the model.
A permitted relation may be:

$$R_{EI} = \{(x_E, x_I) : \kappa(x_E, x_I, b, c) \leq \varepsilon\}.$$

A dual-aspect discrepancy is:

$$\delta_{EI} = \kappa(\hat{x}_E, \hat{x}_I, \hat{b}, c).$$

A large discrepancy may indicate:

- wrong exterior estimate;
- wrong interior estimate;
- wrong coupling model;
- report distortion;
- missing context;
- a genuinely unusual state.

The discrepancy is a diagnostic.
It is not automatic evidence that the person or instrument is dishonest.

## Cross-Domain Causal Questions

Prediction asks:

$$p\left(x_I,\left(t+\tau\right),\mid x_E\left(t\right)x_I\left(t\right)c\right).$$

Causal intervention asks:

$$p\left(x_I,\left(t+\tau\right)\mid do\left(u_E\right)c\right).$$

An exterior-to-interior effect may be:

$$\Delta_{E\to I}=E\left[x_I\mid do\left(u_E=u_1\right)\right]-E\left[x_I\mid do\left(u_E=u_0\right)\right].$$

A modeled interior-to-exterior effect may be:

$$\Delta_{I\to E}=E\left[x_E\mid do\left(u_I=u_1\right)\right]-E\left[x_E\mid do\left(u_I=u_0\right)\right].$$

But an interior intervention must be operationalized.
A command to:
Attend to the breath
is an exterior event.
The resulting interior attention is not guaranteed.
Compliance varies.
Expectation varies.
Therefore the intervention architecture must distinguish:

$$\text{Instruction Assigned}$$

from:

$$\text{Interior Regulation Achieved}.$$

## Mediation

An exterior intervention may affect an interior state through a pathway:

$$u_E\to x_E\to b\to x_I.$$

An interior intention may affect exterior action through:

$$x_I\to b\to y_A.$$

A direct coupling term may be unnecessary if the interface mediates the relationship.
Model comparison should therefore test:

- direct coupling;
- boundary-mediated coupling;
- common cause;
- delayed feedback.

# Temporal Asymmetry

Exterior and interior states may operate on different timescales.
Let:

$$x_E = \left(x_E^{fast}, x_E^{slow}\right), x_I = \left(x_I^{moment}, x_I^{narrative}\right).$$

Fast exterior dynamics may include:

- neural firing;
- movement;
- autonomic response.

Slow exterior dynamics may include:

- endocrine regulation;
- structural plasticity;
- chronic disease;
- institutional change.

Momentary interior dynamics may include:

- perception;
- affect;
- attention.

Slow interior dynamics may include:

- identity;
- purpose;
- autobiographical meaning;
- long-term trust.

A model fitted at one timescale may miss coupling that becomes visible at another.

# Synchronization Is Not Identity

Suppose exterior and interior indicators become temporally aligned.
Define phase difference:

$$\Delta\phi(t) = \phi_E(t) - \phi_I(t).$$

Stable phase relation may support coupling.
It does not establish that the variables are identical.
Two pendulums can synchronize while remaining distinct systems.
One process can track another while preserving a different meaning.

## Exterior Prediction from Interior Variables

A modeled interior state may possess incremental predictive value.

Let exterior-only prediction error be:

$$J_E = E\left[ \| y_E^{t+1} - \hat{y}_E^{t+1}\left(x_E^t\right) \|^2 \right].$$

Let dual-aspect prediction error be:

$$J_{EI} = E\left[ \| y_E^{t+1} - \hat{y}_E^{t+1}\left(x_E^t, x_I^t\right) \|^2 \right].$$

Define improvement:

$$\Delta J = J_E - J_{EI}.$$

If:

$$\Delta J > 0$$

under valid out-of-sample evaluation, interior variables add predictive information.

But this still does not prove independent causal power.

The improvement may arise because the report captures exterior variables not otherwise measured.

## Interior Prediction from Exterior Variables

Similarly:

$$J_I = E\left[ d_I\left(x_I^{t+1}, \hat{x}_I^{t+1}\left(x_I^t\right)\right)\right], J_{IE} = E\left[ d_I\left(x_I^{t+1}, \hat{x}_I^{t+1}\left(x_I^t, x_E^t\right)\right)\right].$$

Improved prediction demonstrates relation.

The form of the relation remains an empirical question.

## Model Comparison

Let candidate models be:

$$M_0, \ldots, M_K.$$

Their predictive scores may be:

$$S_k = -\sum_t \log p\left(D_t, \mid D_{<t} M_k\right).$$

A lower negative log predictive score is better under the declared convention.

Bayesian model evidence may be:

$$p\left(D \mid M_k\right) = \int p\left(D, \mid \theta_k M_k\right) p\left(\theta_k \mid M_k\right) \, d\theta_k.$$

Posterior model probability is:

$$p\left(M_k \mid D\right) \propto p\left(D \mid M_k\right) p\left(M_k\right).$$

But model comparison is relative to the candidates.

The best available model may still be wrong.

## Falsification Conditions

A coupled dual-aspect model should declare what would weaken it.
Possible failure conditions include:

- interior variables add no out-of-sample prediction;
- coupling parameters collapse toward zero;
- exterior-only models perform equally well with fewer assumptions;
- intervention effects contradict the proposed direction;
- report mappings fail across contexts;
- inferred interior states fail to generalize across persons;
- latent coordinates cannot be identified;
- predictions depend upon data leakage;
- replication fails.

A formalism becomes scientific not because it contains equations.
It becomes scientifically useful because the equations reveal where the formalism may fail.

## The Dual-Aspect Uncertainty Signature

Define:

$$U_{DA} = \begin{bmatrix} U_E \\ U_I \\ U_B \\ U_C \\ U_M \\ U_S \end{bmatrix},$$

where:

- $U_E$ is exterior-state uncertainty;
- $U_I$ is interior-state uncertainty;
- $U_B$ is interface uncertainty;
- $U_C$ is coupling uncertainty;
- $U_M$ is model-form uncertainty;
- $U_S$ is semantic or reporting uncertainty.

## Equation Passport 24.17

***Dual-Aspect Uncertainty***

$$U_{DA} = (U_E, U_I, U_B, U_C, U_M, U_S)^T$$

Status: $[ARC/MOD]$

Purpose: Prevent one confidence interval from concealing several different kinds of uncertainty

Possible pattern:

- precise exterior measurement;
- uncertain interior estimate;
- uncertain coupling;
- strong report ambiguity.

Failure condition: All uncertainty is compressed into one scalar confidence score

Architectural significance: Humility becomes mathematical when the model states exactly where it does not know.

# Interior State Spaces Across Persons

A universal interior coordinate system should not be assumed.

For persons $i$ and $j$, report maps may differ:

$$h_R^{(i)} \neq h_R^{(j)}.$$

The same numerical rating may correspond to different uses of a scale.

The same interior condition may be described through different languages.

Context may change the report map:

$$h_R^{(c_1)} \neq h_R^{(c_2)}.$$

Cross-person comparison requires evidence of measurement invariance.

Without it, one cannot assume:

$$r_i = r_j \Rightarrow x_I^{(i)} = x_I^{(j)}.$$

# Cultural and Linguistic State Coordinates

A language may provide distinctions another language does not encode identically.

A culture may organize emotion, selfhood, duty, spirituality, or time differently.

This does not make every interior state incommensurable.

It requires the mapping to be studied rather than assumed.

The formalism may include:

$$\lambda_i$$

for language,

$$c_i$$

for cultural context,

and:

$$h_R^{(\lambda_i, c_i)}.$$

The vocabulary through which experience is reported becomes part of the observation architecture.

## The Interface Includes Relationship

A person may report differently depending upon:

- trust;
- authority;
- stigma;
- fear;
- expected consequence;
- relationship with the listener.

Therefore:

$$b$$

must sometimes include relational variables.

The interview is not an invisible window.

It is an interaction.

The observer becomes part of the measurement context.

## The Ethical Boundary of Interior Inference

Interior-state inference may be used to:

- assist communication;
- improve care;
- understand attention;
- support accessibility;
- detect distress.

It may also be used to:

- manipulate;
- surveil;
- classify;
- deny autonomy;
- infer private belief;
- override testimony;
- punish predicted intention.

A technically accurate model may still be ethically incoherent.

The formalism should therefore attach to every inferred interior state:

$$\hat{x}_I$$

a governance record:

$$\Gamma_I = (\text{purpose}, \text{consent}, \text{uncertainty}, \text{access}, \text{appeal}, \text{retention}, \text{harm}).$$

## The Right to Contest an Interior Model

A person should not be told:

The model knows what you feel better than you do.

without extraordinary evidence, clear purpose, and an opportunity to contest the inference.

At the same time, a person's causal interpretation may be mistaken.

The coherent distinction is:

- the person remains the primary witness to lived experience;
- empirical inquiry remains necessary for mechanism;
- model outputs remain revisable;
- dignity does not depend upon model agreement.

## Dual-Aspect Coherence

Let:

$$C_E(t)$$

represent exterior coherence.

Let:

$$C_I(t)$$

represent interior modeled coherence.

Let:

$$C_B(t)$$

represent interface coherence.

Define:

$$C_{DA} = \begin{bmatrix} C_E \\ C_I \\ C_B \end{bmatrix}.$$

A dual-aspect Cathedral Equation is:

$$\frac{dC_{DA}}{dt} = G_{DA} - L_{DA}.$$

Exterior function may be preserved while interior coherence declines.

Interior purpose may remain while exterior function is impaired.
The interface may fail even when both domains retain organization.
For example:

- a person may possess clear intention but lack motor ability to communicate;
- a person may function outwardly while meaning and hope collapse;
- a person may report confidence while exterior performance deteriorates.

Whole-person assessment should not silently substitute one domain for the others.

## Apparent Agreement

Suppose:

$$C_E \approx C_I.$$

This numerical resemblance does not prove strong coupling unless:

- the variables are valid;
- scales are comparable;
- timing is appropriate;
- common causes are addressed;
- independent evidence exists.

Agreement may arise by construction.
The equations must not discover what the scoring system already forced.

## The Spirit Field Boundary

The Spirit Field remains:

$$S_f[SYM/ARC]$$

as symbolic language for organized interior coherence.
Within Chapter Twenty-Four, it may be related symbolically to:

$$x_I.$$

But:

$$S_f \neq x_I$$

as an empirically established identity.
Nor is:

$$S_f$$

a physical state variable merely because a state-space formalism exists.
A symbolic relation may be:

$$S_f = O_{sym}(x_I, m, v, g, c)[SYM/ARC],$$

where:

- $m$ is memory;
- $v$ is value;
- $g$ is purpose;
- $c$ is compassion or context.

This may support reflection.
It does not become a physical measurement equation.

## What the Dual-Aspect Formalism Claims

It claims that:

- exterior and interior descriptions can be represented distinctly;
- observations must be distinguished from hidden states;
- reports are boundary-mediated evidence;
- state inference requires uncertainty;
- model classes can be compared;
- coupling can be tested;
- observability and identifiability can fail;
- one formal history may admit multiple projections;
- scientific and lived descriptions may constrain one another.

## What the Dual-Aspect Formalism Does Not Yet Claim

It does not yet claim that:

- interior variables are fundamental physical fields;
- the state spaces have been empirically identified;
- one universal mapping joins neural states to experience;
- mental causation has been proven;
- human awareness has been reduced to equations;
- every interior state is observable;
- every report is accurate;
- exterior measurement exhausts lived reality;
- the Spirit Field has become a scientific mechanism;
- the Bridge Species has become a biological taxonomic category.

# Common Failures of Dual-Aspect State-Space Architecture

## State–Observation Collapse

A recorded signal is treated as the hidden state itself.

## Exterior–Environment Collapse

"Exterior" is treated as meaning only what lies outside the body.

## Interior–Spatial Collapse

"Interior" is treated as a newly proposed spatial dimension.

## Latent–Real Collapse

A fitted latent variable is treated as an independently verified entity.

## Coordinate Reification

One mathematically convenient axis is treated as the unique natural organization of the interior.

## Product-Space Innocence

Every exterior state is assumed capable of pairing with every interior state.

## One-to-One Assumption

A unique neural or physiological state is assumed for every lived state without evidence.

## Many-to-One Erasure

Different exterior mechanisms compatible with the same report are ignored.

## Report–Experience Collapse

The words spoken are treated as the entire experience.

## Report Dismissal

A first-person report is rejected merely because it is not an instrument reading.

## Interior Infallibility

A person's causal explanation is treated as necessarily correct because the experience is first-person.

## Exterior Absolutism

Only instrumentally measured variables are treated as real.

## Neural-Correlate Identity

A neural correlate is treated as the complete experience.

## Behavioral Transparency

Observed behavior is treated as uniquely revealing intention or value.

### Silence–Absence Collapse
Failure to report is treated as absence of experience.

### Confidence–Accuracy Collapse
Strong report confidence is treated as proof of historical or causal accuracy.

### Observability Universalism
A state observable under one model and task is assumed observable everywhere.

### Observability–Measurement Collapse
A reconstructible state is described as directly measured.

### Identifiability Neglect
A coupling parameter is interpreted despite multiple parameter sets fitting the same data.

### Practical–Structural Confusion
Mathematical identifiability under ideal data is treated as realistic estimability.

### Controllability–Permission Collapse
Ability to alter an interior or exterior state is treated as ethical authority.

### Coupling–Causality Collapse
A fitted cross-term is treated as proven causal influence.

### Bidirectionality Assumption
Reciprocal coupling is inserted because the framework is called dual-aspect.

### Common-Cause Neglect
A third process capable of generating both descriptions is omitted.

### Readout-Model Neglect
Interior variables are assigned causal power without comparison to a readout model.

### Model-Class Closure
The favored models are treated as exhausting all possible explanations.

### Posterior Certainty
A narrow estimate is trusted despite model misspecification.

### Noise-as-Ignorance Collapse
Every mismatch is assigned to random noise rather than model error.

### Error-as-Deception Collapse
Disagreement between report and exterior measurement is treated as dishonesty.

### Triangulation Double Counting
Several channels derived from one source are treated as independent evidence.

### Context Erasure

Language, culture, trust, and task are omitted from the report model.

### Temporal-Scale Collapse

Fast neural dynamics and long-term meaning are modeled as though they operate on one timescale.

### Takens Overreach

Delay embedding is treated as proof that one scalar signal contains the whole person.

### DCM Overreach

A hidden-neuronal-state model is treated as a model of lived experience without an additional bridge.

### POMDP Anthropomorphism

A belief distribution maintained by a controller is treated as conscious belief.

### State-Sufficiency Overreach

A state sufficient for one prediction is treated as sufficient for every question.

### Minimal-Realization Ontology

A minimal mathematical model is treated as metaphysically complete.

### Cross-Person Invariance Assumption

One report scale is treated as having identical meaning across persons and cultures.

### Privacy Erasure

Inferability is treated as ownership.

### Model Authority

The inferred state is granted more authority than the person, evidence, and uncertainty jointly support.

### Symbolic Physicalization

The Spirit Field is converted into a physical variable merely because interior state spaces are modeled.

## The Dual-Aspect State-Space Constitution

Before any dual-aspect state-space claim enters Part III, it must answer:

1. What system is being modeled?
2. What is the system boundary?
3. What is the time horizon?
4. What is the exterior state space?
5. What is the interior state space?
6. What is the interface state space?
7. Is the joint space a product, subset, manifold, graph, or hybrid?

8. What makes a state admissible?
9. What variables belong to the exterior state?
10. What variables belong to the interior state?
11. What variables belong to the interface?
12. Are any variables counted twice?
13. Is exterior being confused with outside the body?
14. Is interior being confused with physical location?
15. What is the state-transition law?
16. Is time continuous or discrete?
17. Is the model deterministic or stochastic?
18. What inputs are represented?
19. Which inputs are controlled?
20. Which influences are disturbances?
21. What process noise is assumed?
22. What parameters are fixed?
23. What parameters change through time?
24. What observation channels exist?
25. What does each instrument measure?
26. What forward observation map is assumed?
27. What preprocessing occurs?
28. What information is lost through preprocessing?
29. What exterior state remains hidden?
30. What interior state remains hidden?
31. What is directly observed?
32. What is inferred?
33. What is reported?
34. What enables the report?
35. What question elicited the report?
36. What language carries the report?
37. What context shaped the report?
38. What memory process participates?
39. What response bias may exist?
40. What social pressure may exist?
41. Can the person communicate freely?
42. Can silence reflect incapacity or privacy?
43. Is first-person authority preserved?
44. Is first-person fallibility preserved?
45. Is third-person measurement preserved?
46. Are third-person limitations preserved?
47. What mathematical meaning of state is used?
48. Is the state sufficient for the declared prediction?
49. Sufficient over which horizon?
50. Sufficient under which intervention?
51. Is the state Euclidean?
52. Is it constrained?
53. Is it a manifold?
54. Is it categorical?
55. Is distance in the state space meaningful?
56. What metric is used?
57. Are interior variables additive?
58. Are scales comparable?

59. Is the joint state observable?
60. Which state components are observable?
61. Through which sensors?
62. Over which time interval?
63. Under which inputs?
64. What is the observability matrix?
65. What is its rank?
66. Is the Gramian well-conditioned?
67. Is observability local or global?
68. Is observability structural or practical?
69. What state pairs remain indistinguishable?
70. What equivalence classes remain?
71. Is an observer or filter used?
72. What model does the estimator assume?
73. What covariance is assigned?
74. Is model error represented?
75. Can the estimate become overconfident?
76. Is a belief state used?
77. Whose uncertainty does the belief state represent?
78. Is it being confused with lived belief?
79. Is nonlinear observability claimed?
80. What Lie derivatives are used?
81. Does the rank condition hold?
82. Only locally or globally?
83. Is delay reconstruction used?
84. What dynamical assumptions support it?
85. Is the system deterministic?
86. Is the attractor finite-dimensional?
87. Is the observation generic?
88. Is the embedding dimension adequate?
89. Is delay embedding being confused with causal recovery?
90. Is a minimal realization used?
91. What is controllable?
92. What is observable?
93. What dimensions were removed?
94. Are alternative coordinate systems equivalent?
95. Has one latent coordinate been reified?
96. Is the parameter structure identifiable?
97. Could two parameter values generate the same data?
98. Is identifiability structural or practical?
99. What data quantity is required?
100. What sampling rate is required?
101. Is the model exterior-only?
102. Is the interior a readout?
103. Is bidirectional coupling assumed?
104. Is a common latent cause considered?
105. Are boundary-mediated pathways considered?
106. What model classes are compared?
107. What criterion selects among them?
108. Is out-of-sample prediction used?
109. Is intervention evidence used?

110. Is model evidence used?
111. Are complexity penalties used?
112. Are all comparisons relative to the candidate set?
113. What result would falsify the favored model?
114. What result would favor an exterior-only model?
115. What result would favor a readout model?
116. What result would favor a coupled model?
117. What result would favor a common-cause model?
118. Is causal direction claimed?
119. What intervention supports it?
120. Was the intervention actually achieved?
121. Is assigned instruction distinguished from interior compliance?
122. Are expectation effects controlled?
123. Are demand characteristics controlled?
124. What delay connects cause and observation?
125. Are cross-person comparisons made?
126. Is measurement invariance demonstrated?
127. Does language alter the reporting map?
128. Does culture alter the coordinate system?
129. Does trust alter report probability?
130. Is triangulation used?
131. Are evidence channels independent?
132. Are shared biases identified?
133. What happens when channels disagree?
134. Is disagreement treated as information?
135. Is report being confused with causal mechanism?
136. Is absence of measured injury being confused with absence of pain?
137. Is reported pain being confused with proof of one physical cause?
138. What uncertainty belongs to exterior state?
139. What uncertainty belongs to interior state?
140. What uncertainty belongs to the interface?
141. What uncertainty belongs to coupling?
142. What uncertainty belongs to model form?
143. What uncertainty belongs to semantics?
144. Is the interior state inferred for care, research, prediction, or control?
145. Was consent obtained?
146. Who may access the inference?
147. Can the person inspect it?
148. Can the person contest it?
149. Can it be corrected?
150. Can it be deleted?
151. Can it be used for punishment?
152. Can it be used for manipulation?
153. Does inferability exceed ethical permission?
154. Is dignity preserved?
155. Is autonomy preserved?
156. Is privacy preserved?
157. Is the model allowed to override testimony?
158. Under what evidentiary standard?
159. Does the model distinguish experience from explanation?
160. Is dual-aspect coherence being measured?

161. What is exterior coherence?
162. What is interior coherence?
163. What is interface coherence?
164. Can one remain high while another declines?
165. Is one domain being substituted for the whole?
166. Is the joint space being treated as metaphysical proof?
167. Is the formalism compatible with several ontologies?
168. What ontology is actually claimed?
169. Is the Spirit Field invoked?
170. Is its symbolic status explicit?
171. Is any physical-field claim being made?
172. What evidence would be required for such a claim?
173. What remains unobservable?
174. What remains unidentifiable?
175. What remains unreportable?
176. What remains culturally untranslated?
177. What remains ethically private?
178. What would require the model to change?
179. What would require the coupling to be removed?
180. What does the dual-aspect claim not establish?

A dual-aspect claim unable to answer these questions may remain valuable as:

$$[MOD], [ARC], [CONJ], \text{or } [SYM].$$

Its status must remain visible.

## The Architect's Reflection

For many years, I thought the distinction between interior and exterior required choosing sides.

Either reality must be described from the outside.

Or life must be understood from within.

Reality gradually taught another lesson.

The body can be measured.

But the person is lived.

Pain possesses physiology.

Pain also hurts.

A memory possesses exterior conditions.

It also returns through significance.

An intention becomes neural, muscular, linguistic, and social action.

Yet the person does not ordinarily encounter intention as a matrix of measurements.

They encounter:

I will do this.

Perhaps the error was never the existence of two descriptions.

Perhaps the error was requiring one description to humiliate the other.

Exterior science gives humanity extraordinary discipline.

It reveals mechanisms.

It corrects illusion.

It makes treatment, engineering, prediction, and shared verification possible.

Interior testimony reveals what those mechanisms are like within a life.

It reveals suffering.

Hope.

Meaning.

Fear.

Purpose.

Responsibility.

Neither description should become sovereign beyond its evidence.

The instrument should not declare that unmeasured pain is unreal.

The report should not declare that its first causal explanation is complete.

The model should not pretend that a coordinate is the person.

The person should not be asked to disappear so that the equation becomes elegant.

Perhaps the dual aspect is not a war between two realities.

Perhaps it is the discipline of admitting that one reality becomes available through more than one relationship of understanding.

Book I said:

Reality is encountered externally and lived internally.

Part III now adds:

The bridge becomes mathematically responsible only when both sides retain the right to correct the model.

## Quiet Reflection

Pause.

Notice one present interior state.

A sensation.

A memory.

A concern.

A hope.

Do not explain it immediately.

Ask:

What part of this state can I report?

What part remains difficult to place into language?

What exterior changes accompany it?

What exterior measurements would remain silent?

Could another person infer it from my behavior?

Could they infer it incorrectly?

Could I misunderstand its cause while still knowing it is present?

Now consider another person.

How much of their interior life do you know?

How much have you inferred?

How much have you assumed?

What evidence came from their words?

What evidence came from your interpretation?

What remains properly theirs?

Remain with one final question:

How can I become more faithful to another person's lived reality without pretending that I already possess it?

## Constitutional Declaration of Dual-Aspect Integrity

We will distinguish state from observation.

We will distinguish hidden state from measured output.

We will distinguish model from reality.

We will distinguish exterior description from everything merely outside the body.

We will distinguish interior description from physical location.

We will not present the interior state space as a newly discovered spatial dimension.

We will not present it as a separate substance.

We will not present it as a soul coordinate.

We will define every state.

We will define every state space.

We will define every trajectory.

We will define every input.

We will define every disturbance.

We will define every parameter.

We will define every observation.

We will define every report.

We will define every boundary.

We will preserve the interface.

We will remember that the boundary transforms what crosses it.

We will not call a report the experience itself.

We will not call silence absence.

We will not call a neural correlate the whole experience.

We will not call a behavioral pattern a transparent intention.

We will honor first-person authority.

We will preserve first-person fallibility.

We will honor third-person evidence.

We will preserve third-person limitation.

We will not ask either description to erase the other.

We will define observability.

We will declare the sensors.

We will declare the time horizon.

We will declare the admissible inputs.
We will identify indistinguishable states.
We will preserve equivalence classes where evidence cannot identify one state.
We will distinguish observability from transparency.
We will distinguish observability from direct measurement.
We will define identifiability.
We will distinguish state observability from parameter identifiability.
We will distinguish structural identifiability from practical estimability.
We will not interpret a coupling coefficient merely because software produced it.
We will define controllability.
We will not confuse controllability with permission.
We will not confuse reachability with desirability.
We will not use mathematical power to erase consent.
We will preserve uncertainty.
We will preserve process noise.
We will preserve observation noise.
We will preserve report uncertainty.
We will preserve model-form uncertainty.
We will not permit narrow posteriors to conceal wrong models.
We will distinguish the hidden state from the observer's belief state.
We will not call a POMDP belief distribution conscious belief.
We will use nonlinear observability carefully.
We will declare whether conclusions are local or global.
We will use delay reconstruction carefully.
We will not claim that one time series contains the whole person.
We will not claim that reconstructed geometry identifies every cause.
We will preserve coordinate freedom.
We will not reify one latent coordinate system.
We will not confuse a minimal realization with a complete ontology.
We will compare model classes.
We will compare exterior-only models.
We will compare readout models.
We will compare coupled models.
We will compare common-cause models.
We will not insert bidirectional coupling merely because the framework is called dual-aspect.
We will make every coupling earn interpretation.
We will test prediction.
We will test intervention.
We will test generalization.
We will test replication.
We will state what would falsify the model.

We will preserve the possibility that the favored model is wrong.
We will distinguish instruction from achieved interior regulation.
We will examine expectation.
We will examine compliance.
We will examine context.
We will examine language.
We will examine culture.
We will examine trust.
We will not assume one report scale means the same thing for every person.
We will triangulate.
We will not double-count shared evidence.
We will treat disagreement among evidence channels as information.
We will not dismiss pain because one proposed cause was not measured.
We will not infer one cause merely because pain was reported.
We will preserve experience and causal explanation as distinct claims.
We will recognize that exterior states may be hidden.
We will recognize that interior states may be partially reportable.
We will recognize that the experiencing person may also lack complete self-observation.
We will preserve privacy.
We will preserve the right to contest inference.
We will preserve the right to correction.
We will preserve the right to deletion where continued retention lacks justification.
We will not permit inferability to become ownership.
We will not hide moral authority behind a model.
We will preserve exterior coherence.
We will preserve interior coherence.
We will preserve interface coherence.
We will not substitute one for the whole person.
We will preserve one participating history.
We will preserve multiple descriptions.
We will permit formal unity without descriptive erasure.
We will preserve ontological humility.
We will not claim that the joint state space is metaphysical proof.
We will preserve the symbolic status of the Spirit Field.
We will not convert symbolic interior coherence into a physical field merely because mathematics can represent it.
We will remain faithful to Book I's constitutional order.
Reality first.
Observation before interpretation.
Evidence before conclusion.
Science before synthesis.

Mathematics before architectural recognition.
Exterior measurement constrains.
Interior testimony reveals.
The boundary translates.
The model estimates.
Uncertainty remains visible.
Reality decides whether the formalism stands.

## Crossing the Next Arch

The first Formal Laboratory has now established:

- exterior state spaces;
- interior state spaces;
- joint state spaces;
- interface states;
- trajectories;
- observation maps;
- report channels;
- state estimation;
- belief states;
- observability;
- controllability;
- identifiability;
- equivalence classes;
- nonlinear reconstruction;
- coordinate freedom;
- competing model classes;
- first-person data;
- third-person data;
- triangulation;
- causal intervention;
- dual-aspect uncertainty;
- ethical boundaries.

But a state space alone does not explain development.
It tells us what states are possible.
It does not yet tell us how organized patterns:

- recur;
- differentiate repeatedly;
- scale;
- nest;
- preserve ratios;
- change balance;

- bifurcate;
- reproduce structure;
- create new levels of organization.

The next formal question therefore becomes:

How does organized reality repeatedly transform itself across time and scale without requiring exact repetition?

Part III must now distinguish:

- recursion from repetition;
- recursion from fractality;
- exact self-similarity from statistical self-similarity;
- ratios from mystical numerology;
- dynamic balance from fifty–fifty equality;
- scale invariance from superficial resemblance;
- attractors from predestination;
- recursive differentiation from mere complication.

Exterior and interior state spaces have been defined.

The next chapter asks how their relationships transform recursively.

## CATHEDRAL THREAD

Before exterior and interior states could be formalized, Book I had to establish boundary.

Before boundary could become interface, differentiation had to create identity.

Before interface could carry evidence, observation and report had to become distinguishable.

Before hidden states could become scientifically addressable, measurement had to become an observation map.

Before an inferred state could become trustworthy, observability and identifiability had to become visible.

Before partial knowledge could guide action, uncertainty had to become a belief state rather than concealed certainty.

Before first-person data could enter science responsibly, report had to be distinguished from experience and disciplined without being dismissed.

Before exterior measurement could constrain interior interpretation, measured signal had to be distinguished from exterior mechanism.

Before dual-aspect coupling could become a scientific proposal, exterior-only, readout, and common-cause models had to remain genuine competitors.

Before formal unity could become intellectually honest, unity had to be distinguished from identity and descriptive difference from metaphysical separation.

The first principle of Part III is now established:

A state is not what is observed; it is the modeled organization through which observation, history, input, and future evolution become related.

The exterior-state principle follows:

Exterior reality becomes scientifically accessible through observation maps whose limitations remain part of the evidence.

The interior-state principle follows:

Interior state space represents selected dimensions of lived participation without becoming a newly discovered physical dimension or substance.

The interface principle follows:

The boundary organizes the transformation through which exterior events become lived, lived states become action, and experience becomes communicable evidence.

The observability principle follows:

A hidden state becomes distinguishable only through the histories it produces under a declared experiment.

The equivalence principle follows:

Where distinct states produce the same observations, the evidence supports an equivalence class rather than one privileged hidden state.

The estimation principle follows:

State estimates remain conditional upon model, sensor, prior, noise, and context.

The coordinate principle follows:

Latent coordinates may change while observable behavior remains invariant; mathematical convenience must not become ontology.

The first-person principle follows:

Lived report may discipline empirical investigation without becoming infallible or reducible to the report channel.

The triangulation principle follows:

Exterior measurement, first-person report, behavior, and context become strongest when each constrains the others without any one claiming complete authority.

The model-comparison principle follows:

Dual-aspect coupling earns scientific standing only by outperforming serious exterior-only and alternative latent explanations under prediction and intervention.

The ethical principle follows:

The capacity to infer interior states does not grant ownership over another person's interior life.

The dual-aspect principle follows:

One participating history may admit exterior and interior projections whose relations can be modeled without declaring either projection complete.

The constitutional boundary follows:

Two descriptions do not require two universes; one reality does not require one vocabulary.

The recursive principle now approaches:

State spaces define what may be; recursion begins explaining how organized possibility repeatedly becomes.

# EXTERIOR, INTERIOR, AND INTERFACE STATE SPACES



*One reality permits distinct exterior, interior, and interface descriptions.*

**MEASUREMENT != EXPERIENCE | REPORT != WHOLE REALITY**

FS-G21.01 EQ-24.01

### ONE RELATIONAL STATE

Begin from one reality and one relational state before any exterior, interior, or interface projection.

FS-G21.02 EQ-24.02

### EXTERIOR, INTERIOR, INTERFACE SPACES

Declare complementary state spaces without turning them into separate substances or a third hidden entity.

FS-G21.03 EQ-24.06

### EXTERIOR OBSERVATION AND MEASUREMENT

Preserve signal, noise, calibration, measurement error, missing variables, and observation limits.

FS-G21.04 EQ-24.04

### PERSPECTIVE-CONDITIONED READOUT

Make viewpoint, embodiment, context, history, and access conditions explicit.

FS-G21.05 EQ-24.03 · EQ-24.12 · EQ-24.13

### REPRESENTATION, AWARENESS, BOUNDARY

Distinguish representation, access, awareness, threshold conjecture, and symbolic interior language.

FS-G21.06 EQ-24.01 · EQ-24.04

### EMBODIMENT, LEARNING, MEMORY, MEANING

Keep bodily state, affect, learning, memory, valuation, and lived significance jointly relevant but nonidentical.

FS-G21.07 EQ-24.05

### COMPATIBILITY FIBER AND MANY-TO-MANY MAP

Show that one exterior description can support several interior candidates and vice versa.

FS-G21.08 EQ-24.07

### EXPRESSION, REPORT, AND ACTION

Model intention, restraint, action, expression, and report as incomplete, noisy, and noninjective channels.

FS-G21.09 EQ-24.04 · EQ-24.08

### OTHER-PERSON INFERENCE AND EMPATHY

Infer another interior state under uncertainty without equating empathy with direct access.

FS-G21.10 EQ-24.08 · EQ-24.09

### OBSERVABILITY AND EQUIVALENCE

Declare what can be identified, which alternatives remain observationally equivalent, and where inference stops.

FS-G21.11 EQ-24.06 · EQ-24.07 · EQ-24.10

### PRIVACY, CONSENT, NO-REDUCTION

Protect private interiority, consent scope, dignity, shared records, and the no-reduction postulate.

FS-G21.12 EQ-24.10 · EQ-24.11 · EQ-24.12 · EQ-24.13

### RESIDUALS, HOLOFRACTAL, REVISION

Track compatibility residuals, holofractal candidates, conjectural thresholds, Spirit Field boundaries, and correction.

**DIGNITY | UNCERTAINTY | SOURCE TRANSPARENCY | FAILURE | REVISION**

FS identifies figure position; EQ identifies canonical equation. Reality remains the measure.

CHAPTER TWENTY-FIVE

# RECURSION, FRACTALS, RATIOS, AND DYNAMIC BALANCE

*Scale, Self-Similarity, Recursive Differentiation, and Organized Proportion*

*The Second Formal Laboratory*

**Reality rarely repeats itself exactly.**

**It carries relationships forward, transforms them under new conditions, and becomes capable of something new.**

**Balance is not equality of contribution.**

**Balance is relationship held in sufficient proportion for continued becoming.**

## Guiding Question

How does organized reality repeatedly transform itself across time and scale without requiring exact repetition; how can recursion, iteration, fixed points, bifurcations, chaos, self-similarity, fractal dimension, scale transformations, ratios, and dynamic balance be formalized; and what prevents structural resemblance across physics, biology, consciousness, culture, and civilization from being misidentified as one universal mechanism?

Chapter Twenty-Four established:

- exterior state spaces;
- interior state spaces;
- interface states;
- observation maps;
- hidden-state estimation;
- observability;
- identifiability;
- competing model classes;
- dual-aspect uncertainty.

State spaces tell us what a formal system may become.

They do not yet tell us how becoming proceeds.

A present state may be transformed into a later state.

A later state may become the input to another transformation.

A pattern may recur.

A boundary may be inherited.

A ratio may remain approximately stable.

A process may branch.

A scale may change.

A system may return toward a prior organization.

A small difference may be amplified until two trajectories become unrecognizable.

Simple rules may produce fixed points, cycles, bifurcations, or deterministic chaos.

A form may appear self-similar across a range of scales.

But these possibilities must remain mathematically distinguished.

Book I already established the architectural recognition of **recursive differentiation**:

Reality rarely differentiates only once. Each new level inherits previous organization while introducing genuinely new capacities.

It also insisted that recursive differentiation does more than increase complexity: it generates history, because each differentiated system begins a distinctive process of becoming.

The Project Cathedral charter gives Part III the governing rule:

**THE PATTERN MAY RECUR. THE MECHANISM CHANGES.**

It explicitly requires cross-scale similarity to be treated as **structural analogy rather than identity of mechanism**, while distinguishing mathematical architectures such as ratios, recursion, scale maps, and dynamical systems from SDAE formal models and symbolic ontological interpretations.

A second inheritance is equally important:

**Balance does not mean 50/50. Balance means the right relationship for stability, dignity, and renewal.**

Within SDAE's ancestral-symbolic language, Ma'at names right relationship rather than a physical force or hidden mechanism.

Chapter Twenty-Five must now formalize all three recognitions:

Inheritance → Recursive Transformation → Differentiation → Cross-Scale Relation → Dynamic Rebalancing *[ARC]*

The central scientific discipline remains:

$$P_{n+1} \sim P_n \text{ does not imply } M_{n+1} = M_n$$

where:

- $P_n$ is an observed or modeled pattern at level $n$;
- $M_n$ is the mechanism generating it;
- $\sim$ means a declared form of structural resemblance;
- ¿means identity.

The chapter therefore asks not merely:

Does a pattern repeat?

It asks:

**What is preserved, what is transformed, what becomes newly possible, over which scales does resemblance hold, and what mechanism operates at each level?**

# Terminological Discipline

| Term | Meaning in Part III |
|---|---|
| **Recursion** | Construction or update in which a later object is defined partly through an earlier object of the same formal sequence |
| **Iteration** | Repeated application of a rule or map |
| **Recurrence relation** | Equation relating a sequence element to one or more preceding elements |
| **Recursive differentiation** | Repeated transformation through which inherited organization becomes further distinguished while retaining selected prior constraints |
| **Recursive depth** | Number of successive update, branching, or scale operations applied |
| **Orbit** | Sequence of states generated by repeated application of a map |
| **Fixed point** | State unchanged by one application of a map |

| Term | Meaning in Part III |
|---|---|
| **Periodic orbit** | Sequence returning to the same state after a finite number of iterations |
| **Attractor** | State or set approached by nearby trajectories under declared dynamics |
| **Basin of attraction** | Initial states whose trajectories approach a specified attractor |
| **Bifurcation** | Qualitative change in a system's long-term dynamics as a parameter changes |
| **Chaos** | Deterministic dynamics exhibiting properties such as sensitive dependence, aperiodicity, and bounded complex behavior under a declared definition |
| **Lyapunov exponent** | Average exponential rate at which nearby trajectories converge or diverge |
| **Recurrence** | Return of a state or pattern near an earlier state or pattern |
| **Exact recurrence** | Return to precisely the same modeled state |
| **Approximate recurrence** | Return within a declared tolerance |
| **Structural recurrence** | Reappearance of selected relational features despite differences in state, scale, material, or mechanism |
| **Self-similarity** | Resemblance of an object or process to transformed portions of itself |
| **Exact self-similarity** | Exact reconstruction through scaled copies under declared maps |
| **Statistical self-similarity** | Preservation of probability distributions or statistical scaling properties rather than exact geometry |
| **Affine self-similarity** | Resemblance under transformations permitting unequal scaling in different directions |
| **Scale invariance** | Preservation of a declared mathematical relation under change of scale |
| **Scaling law** | Relation in which one quantity varies as a power or another declared function of scale |
| **Fractal** | Mathematical or empirical object exhibiting specified forms of irregularity, scaling, or noninteger dimension under a declared definition |
| **Fractal dimension** | Dimension-like quantity measuring how detail or occupied structure changes with scale |
| **Box-counting dimension** | Scaling exponent relating the number of covering boxes to box size |
| **Hausdorff dimension** | Measure-theoretic dimension defined through Hausdorff measure |
| **Similarity dimension** | Exponent determined by the contraction ratios of a self-similar construction |
| **Multifractal** | Object requiring a spectrum of local scaling exponents rather than one global exponent |
| **Coarse-graining** | Mapping from finer description to a lower-dimensional or larger-scale description |
| **Scale map** | Rule relating descriptions at two scales |
| **Renormalization** | Domain-specific transformation of variables, couplings, or probability descriptions under change of scale |
| **Renormalization fixed point** | Model or parameter configuration invariant under a declared scale transformation |
| **Universality** | Shared macroscopic or scaling behavior among distinct systems within a defined mathematical class |
| **Ratio** | Quotient comparing two quantities |

| Term | Meaning in Part III |
|---|---|
| **Dimensionless ratio** | Ratio in which physical units cancel |
| **Proportion** | Equality or organized relation among ratios |
| **Composition** | Vector of nonnegative component shares summing to one |
| **Golden ratio** | Positive solution $\varphi$ of $\varphi^2 = \varphi + 1$ |
| **Dynamic balance** | Bounded or stationary organization maintained through continuing, potentially unequal processes |
| **Static equality** | Numerical equality at one time |
| **Point balance** | Instantaneous equality of declared net generation and loss |
| **Cycle balance** | Zero net change over a complete cycle despite changing instantaneous rates |
| **Steady state** | State whose declared macroscopic variables remain constant while internal flows may continue |
| **Equilibrium** | State satisfying the relevant balance conditions of a specified scientific theory |
| **Flux balance** | Condition in which incoming, outgoing, production, and consumption rates yield zero net accumulation |
| **Balance manifold** | Set of states satisfying declared balance equations |
| **Target ratio** | Context-dependent proportion judged viable or desirable for a specified process |
| **Ratio error** | Deviation between an observed ratio and its declared target |
| **Organized proportion** | Unequal or equal contributions arranged sufficiently to preserve the declared system or purpose |
| **Scale commutation** | Agreement between evolving then coarse-graining and coarse-graining then evolving |
| **Commutation defect** | Quantified disagreement between those two procedures |
| **Pattern mechanism distinction** | Rule that structural similarity does not establish common causal generation |
| **Holofractal** | SDAE architectural language for recurring part–whole and cross-scale organization; not, by itself, proof of exact mathematical fractality or holography |

These distinctions must remain visible.

$$\text{Recursion} \neq \text{Repetition} \quad \text{Iteration} \neq \text{Learning} \quad \text{Recurrence} \neq \text{Identity}$$
$$\text{Recursion} \neq \text{Fractality} \quad \text{Visual Complexity} \neq \text{Fractal Scaling}$$
$$\text{Self-Similarity} \neq \text{Whole-in-Every-Part}$$
$$\text{Scale Invariance} \neq \text{One Mechanism at Every Scale}$$
$$\text{Universality} \neq \text{Universal Ontology} \quad \text{Chaos} \neq \text{Randomness}$$
$$\text{Golden Ratio} \neq \text{Universal Law of Beauty or Health}$$
$$\text{Balance} \neq \text{Equal Shares} \quad \text{Steady State} \neq \text{Absence of Activity}$$

## The Historical Inheritance of Ratio

Before ratio became an algebraic symbol, it was embodied in:

- land measurement;
- construction;
- slope;
- volume;
- division;
- astronomy;
- architecture;
- exchange.

The surviving Rhind Mathematical Papyrus provides direct evidence of mathematical instruction in Kemet. The British Museum describes the papyrus as a probable textbook through which scribes learned to solve mathematical problems. Its pyramid problems include the **seked**, a relational measure of a pyramid face's horizontal displacement per unit vertical rise. This is not modern trigonometry written in modern notation; it is nevertheless a disciplined use of proportion to govern built geometry.

The appropriate historical conclusion is neither diminishment nor fabrication.

It is:

**African mathematical inheritance contains surviving evidence of practical ratio, measure, geometry, and architectural proportion long before modern symbolic algebra.**

This chapter does not use the Rhind Papyrus to claim that ancient scribes possessed:

- modern dynamical-systems theory;
- Hausdorff dimension;
- renormalization groups;
- chaos theory;
- the contemporary golden-ratio vocabulary;
- SDAE equations.

The inheritance is more fundamental.

The builders understood that form depends upon measure and that measure frequently depends upon relationship.

Euclid's later geometric tradition formalized what Book VI calls cutting a line in **extreme and mean ratio**, the relation now associated with the golden ratio.

The chapter therefore receives two historical lessons:

Ratio first appears as relation made measurable.
Proof later makes the relation formally transferable.

No unsupported claim of direct transmission is required.

Humanity inherits from every faithfully preserved mathematical stone.

## Scientific and Mathematical Foundations

This laboratory stands upon the inheritance of:

- recurrence relations;
- difference equations;
- differential equations;
- dynamical systems;
- nonlinear dynamics;
- chaos theory;
- fractal geometry;
- measure theory;
- iterated function systems;
- probability;
- statistical mechanics;
- renormalization;
- control theory;
- network science;
- compositional mathematics.

Robert May demonstrated that simple deterministic first-order difference equations can produce fixed points, oscillations, period-doubling cascades, and highly complicated aperiodic dynamics. His analysis helped make clear that mathematical simplicity of the update rule does not imply simplicity of long-term behavior.

Edward Lorenz showed that a small deterministic system of nonlinear differential equations could generate bounded, nonperiodic behavior whose trajectories were highly sensitive to small differences in initial state. Determinism therefore does not guarantee practical long-range predictability.

Mitchell Feigenbaum identified quantitative scaling regularities shared by a broad mathematical class of nonlinear maps undergoing period-doubling transitions. His result is a disciplined example of **universality**: different functions can share scaling structure without becoming the same function or representing the same physical mechanism.

Benoit Mandelbrot used statistical self-similarity and fractional dimension to analyze irregular geographical curves whose measured length changes with measuring scale. His work showed that dimension need not be restricted to the ordinary integer assigned to a smooth line, surface, or volume.

John Hutchinson formalized self-similar sets generated by finite families of contraction maps. Under the relevant assumptions, the associated set operator possesses a unique nonempty compact fixed set, giving a rigorous foundation for iterated-function-system fractals.

Kenneth Wilson and John Kogut's renormalization-group treatment showed how systematic changes of scale can transform parameters and descriptions while revealing fixed points and universal critical behavior. This is a domain-specific mathematical and physical framework—not permission to claim that every biological, psychological, or civilizational scale is governed by one renormalization mechanism.

These inheritances lead to the second formal recognition of Part III:

**A rule may be simple while its history becomes complex.**

**A pattern may recur while the state changes.**

**A scaling relation may be universal within a mathematical class while the mechanisms remain different.**

## Architectural Recognition

Only after the historical, scientific, and mathematical inheritances have spoken does the Architecture introduce its central recognition:

**Organized becoming is often recursive without being repetitious: later states inherit constraints from earlier organization, transform them under new conditions, and may preserve relational motifs while generating genuinely new capacities.**

A general recursive architecture is:

$$R_{n+1} = F_n\left(R_n, U_n, E_n, H_n\right)\left[ARC/MOD\right]$$

where:

- $R_n$ is relational organization at recursive depth, developmental stage, or descriptive level $n$;
- $U_n$ is an input or update condition;

- $E_n$ is environment or context;
- $H_n$ is inherited history;
- $F_n$ is the transformation applied at that level.

If the same map acts at every step:

$$R_{n+1} = F(R_n),$$

the recursion is autonomous.

If the map changes:

$$F_n \neq F_{n+1},$$

the recursion remains historically recursive but not rule-identical.

This distinction is essential for SDAE.

Reality may preserve:

- boundary;
- feedback;
- branching;
- ratio;
- differentiation;
- integration;
- repair;

without preserving the same material mechanism.

## Recursion

A recursion begins with an initial condition:

$$x_0 \in X$$

and an update:

$$x_{n+1} = F(x_n).$$

Repeated substitution gives:

$$x_1 = F(x_0), x_2 = F(F(x_0)) = F^2(x_0),$$

and generally:

$$x_n = F^n(x_0).$$

The sequence:

$$O_F(x_0) = (x_0, x_1, x_2, \ldots)$$

is the orbit of $x_0$.

## Equation Passport 25.1

*General Recursive Update*

$$x_{n+1} = F_n(x_n, u_n)$$

**Status:** $[¿/MOD]$

**Domain:**

$$F_n : X \times U \to X$$

**State:** $x_n$

**Input:** $u_n$

**Update rule:** $F_n$

**Recursive depth:** $n$

**Units:** Determined by the state and input variables

**Autonomous form:**

$$x_{n+1} = F(x_n)$$

**History-dependent form:**

$$x_{n+1} = F(x_n, x_{n-1}, \ldots, u_n)$$

**Required declarations:**

- state space;

- initial state;
- update rule;
- parameter values;
- time interpretation;
- noise;
- boundary constraints.

**What recursion establishes:** Dependence of later organization upon earlier organization

**What it does not establish:**

- exact repetition;
- fractality;
- improvement;
- consciousness;
- progress;
- one mechanism across scales.

**Architectural significance:** The future does not begin without inheritance; it begins by transforming what has already become.

## Recursion Is Not Repetition

Exact repetition requires:

$$x_{n+1} = x_n.$$

Recursion requires only:

$$x_{n+1} = F(x_n).$$

The later state may be:

- larger;
- smaller;
- differentiated;
- reversed;
- stabilized;
- destabilized;
- branched;
- compressed;
- enriched.

A seed becoming a tree is not the tree repeating the seed.

A student developing understanding is not the student repeating the lesson unchanged.

A civilization inheriting mathematics is not merely copying every prior notation.

The transformation carries inheritance forward.

It need not preserve form exactly.

## Recursive Composition

For time-varying maps:

$$x_{n+1}=F_n(x_n),$$

the state after $N$ steps is:

$$x_N=F_{N-1}\circ F_{N-2}\circ\cdots\circ F_0(x_0).$$

The order matters.

Generally:

$$F_i\circ F_j\neq F_j\circ F_i.$$

A system exposed to injury and then repair may not become identical to one repaired before injury.

A person taught one concept before another may form a different conceptual architecture.

Historical order participates in becoming.

Recursive systems are often path-dependent.

## Fixed Points

A fixed point satisfies:

$$x^{¿}=F(x^{¿}).$$

Applying the map again changes nothing:

$$F^{n}\left(x^{¿}\right)=x^{¿}.$$

For a one-dimensional differentiable map, local stability is determined by:

$$\left| F^{'}\left(x^{¿}\right) \right|.$$

If:

$$\left| F^{'}\left(x^{¿}\right) \right| < 1,$$

nearby states contract toward the fixed point.

If:

$$\left| F^{'}\left(x^{¿}\right) \right| > 1,$$

nearby states move away.

The boundary case:

$$\left| F^{'}\left(x^{¿}\right) \right| = 1$$

requires further analysis.

## Equation Passport 25.2

***Fixed Point and Local Stability***

$$x^{¿}=F\left(x^{¿}\right)\delta_{n+1}\approx F^{'}\left(x^{¿}\right)\delta_{n}$$

**Status:** $\left[THM / MOD\right]$

**Perturbation:**

$$\delta_{n}=x_{n}-x^{¿}$$

**Locally attracting:**

$$\left| F^{'}\left(x^{¿}\right) \right| < 1$$

**Locally repelling:**

$$|F'(x^{*})|>1$$

**Required assumptions:**

- differentiable one-dimensional map;
- sufficiently small perturbation;
- neighborhood in which linearization is informative.

**What a fixed point does not imply:**

- physical immobility;
- absence of internal flow;
- desirability;
- whole-system coherence.

**Architectural significance:** A recursive process may discover an organization capable of reproducing its own relational condition.

## Periodic Orbits

A point belongs to a period-$p$ orbit when:

$$F^{p}(x)=x$$

and no smaller positive integer $k<p$ satisfies:

$$F^{k}(x)=x.$$

A two-cycle satisfies:

$$x_2=F(x_1),x_1=F(x_2),$$

with:

$$x_1\neq x_2.$$

The system is recurrent.

It is not static.

This becomes important for dynamic balance.

A coherent rhythm may require alternation:

- contraction and expansion;
- wake and sleep;
- excitation and inhibition;
- speaking and listening;
- activity and recovery.

## The Logistic Map

Consider:

$$x_{n+1} = r\,x_n(1 - x_n),$$

with:

$$0 \le x_n \le 1, 0 \le r \le 4.$$

The model has fixed points:

$$x_0^{¿} = 0$$

and, when $r > 1$,

$$x_1^{¿} = 1 - \frac{1}{r}.$$

The derivative is:

$$F'(x) = r(1 - 2x).$$

At the nonzero fixed point:

$$F'(x_1^{¿}) = 2 - r.$$

Thus the nonzero fixed point is locally stable when:

$$|\,2 - r\,| < 1,$$

or:

$$1 < r < 3.$$

As $r$ increases beyond 3, the fixed point loses stability and periodic behavior emerges. Further parameter changes produce repeated period-doubling, eventually entering regions of deterministic chaos interrupted by periodic windows. May used this and related maps to show how simple nonlinear recurrence can produce remarkably complicated dynamics.

## Equation Passport 25.3

*Logistic Recursion*

$$x_{n+1} = r\, x_n (1 - x_n)$$

**Status:** $[MOD]$

**State:** $x_n$

**Control parameter:** $r$

**State range in the standard normalized map:**

$$x_n \in [0,1]$$

**Parameter range preserving the interval:**

$$r \in [0,4]$$

**Nonzero fixed point:**

$$x^{*} = 1 - \frac{1}{r}$$

**Local-stability range:**

$$1 < r < 3$$

**Possible regimes:**

- extinction or zero attraction;
- stable nonzero fixed point;
- periodic orbit;
- period-doubling;
- chaotic dynamics;
- periodic windows.

**What it demonstrates:** Qualitative behavior can change dramatically while the equation remains unchanged

**What it does not establish:**

- that every population follows the logistic map;
- that every recursive system becomes chaotic;
- that chaos is metaphysical creativity;
- that one parameter explains biological history.

**Architectural significance:** The same rule may support multiple worlds of behavior depending upon relational proportion.

## A Numerical Walk Through Recursion

Choose:

$$x_0 = 0.2.$$

For:

$$r = 2.8,$$

the orbit converges toward:

$$x^{*} = 1 - \frac{1}{2.8} \approx 0.642857.$$

For:

$$r = 3.2,$$

the long-term orbit alternates approximately between:

$$0.513045$$

and:

$$0.799455.$$

For:

$$r = 3.5,$$

the orbit approaches a four-cycle with approximate values:

$$0.382820, 0.826941, 0.500884, 0.874997.$$

For:

$$r = 3.9,$$

the trajectory becomes aperiodic and highly sensitive to its initial condition.

The formal lesson is not that increasing $r$ means increasing disorder in every domain.

It is:

**A change in one relational parameter can reorganize the entire long-term structure of a recursive system.**

## Bifurcation

Let a system depend upon parameter $\mu$:

$$x_{n+1} = F(x_n; \mu).$$

A bifurcation occurs when a small change in $\mu$ produces a qualitative change in the attractor or stability structure.

Possible changes include:

- appearance of a new fixed point;
- loss of stability;
- emergence of a periodic orbit;
- merging of attractors;
- transition into chaos;
- crisis or disappearance of an attractor.

A bifurcation is not merely a large numerical change.

It is a change in the organization of possible trajectories.

## Chaos Is Not Randomness

A chaotic system may be completely deterministic:

$$x_{n+1} = F\left(x_n\right).$$

The next state is fixed by the present state.

Yet small uncertainty in the initial state may grow rapidly.

Suppose:

$$x_0$$

and:

$$x_0 + \delta_0$$

are close.

After $\eta$ steps:

$$|\,\delta_n\,| \approx |\,\delta_0\,|\,e^{\lambda n}.$$

If:

$$\lambda > 0,$$

nearby trajectories separate exponentially on average.

Lorenz's work established that bounded deterministic nonlinear systems can produce nonperiodic trajectories highly sensitive to small initial differences.

$$\text{Deterministic} \not\Rightarrow \text{Predictable at Every Horizon}.$$
$$\text{Unpredictable at Long Horizon} \not\Rightarrow \text{Randomly Generated}.$$

## Lyapunov Exponents

For a one-dimensional differentiable map:

$$x_{n+1} = F\left(x_n\right),$$

the maximal Lyapunov exponent along an orbit is:

$$\lambda = \lim_{N \to \infty} \frac{1}{N} \sum_{n=0}^{N-1} \ln |F'(x_n)|,$$

when the limit exists.

Interpretation:

$$\lambda < 0$$

indicates average local contraction.

$$\lambda = 0$$

often accompanies a transition or neutral direction.

$$\lambda > 0$$

indicates average exponential separation.

## Equation Passport 25.4

*Lyapunov Exponent*

$$\lambda = \lim_{N \to \infty} \frac{1}{N} \sum_{n=0}^{N-1} \ln |F'(x_n)|$$

**Status:** [¿/*MOD*]

**Dynamics:** $F$

**Orbit:** $x_n$

**Units:** In discrete time, logarithmic separation per iteration

**Interpretation:**

- negative: average convergence;
- zero: neutral or critical behavior;
- positive: sensitive dependence.

**Required declarations:**

- norm;
- orbit;

- sampling interval;
- transient removal;
- finite-data estimation method.

**Failure conditions:**

- insufficient data;
- observation noise;
- nonstationarity;
- wrong embedding;
- multiple exponents compressed into one;
- transient behavior mistaken for asymptotic dynamics.

**Architectural significance:** Vulnerability to small difference can be quantified without converting unpredictability into metaphysical randomness.

## Universality Without Identity

Let:

$$\mu_1, \mu_2, \ldots$$

be successive parameter values at which period-doubling bifurcations occur.

Define:

$$\delta_n = \frac{\mu_n - \mu_{n-1}}{\mu_{n+1} - \mu_n}.$$

For a broad class of smooth unimodal maps, Feigenbaum showed that:

$$\delta_n \to \delta \approx 4.6692016\ldots$$

under the relevant assumptions.

The striking fact is that different nonlinear maps can share the same limiting scaling constant even though their formulas differ.

### Equation Passport 25.5

*Period-Doubling Universality*

$$\delta = \frac{\lim_{n \to \infty} \mu_n - \mu_{n-1}}{\mu_{n+1} - \mu_n}$$

**Status:** $[THM / MOD]$

**Bifurcation parameters:** $\mu_n$

**Universal constant for the specified map class:**

$$\delta \approx 4.6692016$$

**Required mathematical class:**

- smooth unimodal maps;
- appropriate quadratic maximum;
- period-doubling route;
- stated regularity conditions.

**What universality establishes:** Shared scaling behavior within a defined class

**What it does not establish:**

- identical mechanisms;
- identical state variables;
- one universal equation of nature;
- common metaphysical meaning.

**Architectural significance:** Deep relational similarity can survive substantial difference in local form.

## Recurrence Across Scale

Temporal recursion asks:

$$x_{n+1} = F(x_n).$$

Scale recursion asks:

$$x^{(l+1)} = S_l(x^{(l)}),$$

where:

- $l$ indexes descriptive scale;
- $S_l$ is a coarse-graining, aggregation, rescaling, or reconstruction map.

Examples of scale change include:

- molecules to cellular variables;
- neurons to population activity;
- individuals to institutional measures;
- fine geometry to coarse geometry;
- microscopic couplings to effective macroscopic parameters.

The map:

$$S_l$$

is not necessarily invertible.

Information may be lost.

Several fine states may produce the same coarse state.

Scale Change Is Often Many-to-One.

## Renormalization

Let:

$$g$$

represent a vector of model parameters or couplings.

A scale transformation is:

$$g' = R_b(g),$$

where $b > 1$ is the scale factor.

A renormalization fixed point satisfies:

$$g^{¿} = R_b(g^{¿}).$$

Near the fixed point:

$$g' - g^{¿} \approx B\left(g - g^{¿}\right),$$

where:

$$B = D\,R_b\left(g^{¿}\right)$$

is the linearized scale-transformation operator.

Its eigenvalues classify directions as:

- relevant;
- irrelevant;
- marginal;

under the domain's definitions.

Wilson and Kogut's renormalization framework explains how systems with different microscopic details may approach common large-scale behavior near critical points.

## Equation Passport 25.6

***Scale Transformation and Fixed Point***

$$g' = R_b(g)\,g^{¿} = R_b\left(g^{¿}\right)$$

**Status:** $[THM / MOD]$

**Parameter space:** $G$

**Scale factor:** $b$

**Scale map:** $R_b$

**Fixed point:** $g^{¿}$

**Linearization:**

$$g' - g^{¿} \approx D\,R_b\left(g^{¿}\right)\left(g - g^{¿}\right)$$

**What it can reveal:**

- critical behavior;
- scale invariance;

- relevant variables;
- universality classes;
- effective descriptions.

**What it does not authorize:**

- applying one physical renormalization group to every domain;
- treating psychology as coarse-grained quantum field theory;
- identifying cultural recursion with critical phenomena;
- claiming universal scale invariance from visual resemblance.

**Architectural significance:** Some details disappear under scale change while selected relational constraints become increasingly dominant.

## Pattern Recurrence, Mechanism Change

The governing Project Cathedral rule can now be formalized.

Let:

$$P_l$$

be the pattern at scale $l$.

Let:

$$M_l$$

be the generating mechanism.

Let:

$$\Phi_l(P_l)$$

be a declared feature representation.

Let:

$$T_l$$

be an admissible transformation between feature spaces.

Define structural discrepancy:

$$d_{str}(l,l+1)=\inf_{T_l\in T_l}\left\|\Phi_{l+1}(P_{l+1})-T_l\Phi_l(P_l)\right\|.$$

Define mechanism discrepancy:

$$d_{mech}(l,l+1)=d_M(M_{l+1},M_l).$$

Structural recurrence may be declared when:

$$d_{str}\le\varepsilon_{str},$$

while mechanism identity would require:

$$d_{mech}=0$$

under an appropriate metric.

The SDAE condition is usually:

$$d_{str}\ \text{small}, d_{mech}\ \text{nonzero}.$$

## Equation Passport 25.7

***Structural Recurrence with Mechanism Difference***

$$P_{l+1}\sim P_l$$

while:

$$M_{l+1}\neq M_l.$$

**Expanded model:**

$$d_{str}=\inf_T\ \|\ \Phi_{l+1}(P_{l+1})-T\,\Phi_l(P_l)\ \|\ d_{mech}=d_M(M_{l+1},M_l)$$

**Status:** $[ARC/MOD/CONJ]$

**Structural feature map:** $\Phi$

**Scale transformation:** $T$

**Pattern tolerance:** $\varepsilon_{str}$

**Mechanism model:** $M_l$

**What it permits:** Quantified comparison without claiming identity

**What it requires:**

- declared features;
- declared scale map;
- declared mechanism representation;
- null comparisons;
- uncertainty;
- alternative explanations.

**Architectural significance:** The pattern may recur. The mechanism changes.

## A Recurrence Score

A bounded similarity score may be defined:

$$R_{l,l+1} = \exp\left[\frac{-d_{str}^2}{\sigma_{str}^2}\right].$$

Then:

$$0 < R_{l,l+1} \leq 1.$$

A value near $1$ indicates strong structural similarity under the declared feature map.

A value near $0$ indicates weak similarity.

This score is model-dependent.

Changing:

- features;
- metric;
- normalization;
- scale map;

may change the score.

It therefore cannot be treated as an observer-independent measure of cosmic resemblance.

## Self-Similarity

An exact self-similar set $A$ may satisfy:

$$A = ¿ i=1 ¿ N\, w_i(A),$$

where each:

$$w_i : X \to X$$

is a contraction or similarity map under the declared construction.

Define the Hutchinson operator:

$$H(B) = ¿ i=1 ¿ N\, w_i(B).$$

The self-similar set is a fixed point:

$$A = H(A).$$

Under Hutchinson's contractive iterated-function-system assumptions, there is a unique nonempty compact fixed set approached through repeated application of the set operator.

### Equation Passport 25.8

*Iterated Function System*

$$W = (w_1, \ldots, w_N)\, H(B) = ¿ i=1 ¿ N\, w_i(B)\, A = H(A)$$

**Status:** $[THM / MOD]$

**Metric space:** $X$

**Contractions:** $w_i$

**Set operator:** $H$

**Attractor:** $A$

**Recursive construction:**

$$A_n = H^n(A_0)$$

**Required assumptions:**

- complete metric space;
- finite family of contractions;
- nonempty compact starting set.

**What it establishes:** Exact mathematical self-similarity of the generated attractor

**What it does not establish:**

- that a natural object is exactly fractal;
- that the construction is physically realized to infinite scale;
- that every part contains the whole;
- that one IFS describes biological development.

**Architectural significance:** Repeated local transformations can generate a globally organized form whose rule is present throughout its construction.

## Similarity Dimension

Suppose each map has contraction ratio:

$$r_i \in (0,1).$$

The similarity dimension $D$ solves:

$$\sum_{i=1}^{N} r_i^D = 1.$$

If every ratio is equal:

$$r_i = r,$$

then:

$$N r^D = 1,$$

so:

$$D = \frac{\ln N}{\ln(1/r)}.$$

Under suitable separation conditions, this dimension agrees with the Hausdorff dimension of the self-similar set.

## Equation Passport 25.9

*Similarity Dimension*

$$\sum_{i=1}^{N} r_i^D = 1$$

**Status:** $[THM / MOD]$

**Number of pieces:** $N$

**Contraction ratios:** $r_i$

**Similarity dimension:** $D$

**Equal-ratio form:**

$$D = \frac{\ln N}{\ln(1/r)}$$

**Required caution:** Equality with Hausdorff dimension requires appropriate separation assumptions

**What dimension expresses:** How multiplicity competes with contraction across scale

**What it does not express automatically:**

- complexity in every sense;
- biological health;
- consciousness;
- moral worth;
- information completeness.

**Architectural significance:** Dimension may become a relation between how many differentiated pieces appear and how strongly each piece is scaled.

# Three Exact Examples

## Cantor Set

Two copies, each scaled by:

$$r = \frac{1}{3}.$$

Therefore:

$$D_C = \frac{\ln 2}{\ln 3} \approx 0.63093.$$

### Koch Curve

Four copies, each scaled by:

$$r = \frac{1}{3}.$$

Therefore:

$$D_K = \frac{\ln 4}{\ln 3} \approx 1.26186.$$

### Sierpiński Triangle

Three copies, each scaled by:

$$r = \frac{1}{2}.$$

Therefore:

$$D_S = \frac{\ln 3}{\ln 2} \approx 1.58496.$$

The dimensions fall between familiar integers.

The Koch curve is more spatially extensive than an ordinary smooth curve but does not fill a two-dimensional region.

The Sierpiński triangle occupies more relational structure than a one-dimensional line but less than an ordinary filled surface.

## Box-Counting Dimension

Let:

$$N(\varepsilon)$$

be the minimum number of boxes of side length $\varepsilon$ required to cover a set.

The box-counting dimension is:

$$D_B = \frac{\lim_{\varepsilon \to 0} \ln N(\varepsilon)}{\ln(1/\varepsilon)},$$

when the limit exists.

Equivalently, a scaling law may appear:

$$N(\varepsilon) \propto \varepsilon^{-D_B}.$$

Taking logarithms:

$$\ln N(\varepsilon) \approx D_B \ln(1/\varepsilon) + c.$$

A straight region on a log–log plot may provide evidence of scaling over a specified range.

It does not prove infinite self-similarity.

## Statistical Self-Similarity

Many empirical structures are not exact copies of themselves.

Instead, statistical properties may scale.

A stochastic process $X(t)$ may satisfy:

$$\{X(at)\}\{a^H X(t)\},$$

where:

- denotes equality in distribution;
- $H$ is a scaling exponent.

Mandelbrot's coastline analysis used statistical self-similarity and fractional dimension rather than claiming that each coastal segment is an exact miniature of the entire coast.

This distinction is central.

$$\text{Exact Geometry} \neq \text{Statistical Scaling}.$$

## Finite Scaling Ranges

No empirical measurement reaches:

$$\varepsilon = 0.$$

A natural object may exhibit apparent scaling only for:

$$\varepsilon_{min} \leq \varepsilon \leq \varepsilon_{max}.$$

Outside that interval:

- molecular discreteness may appear;
- organism size may impose a limit;
- measurement resolution may fail;
- another mechanism may dominate;
- finite boundaries may interrupt scaling.

An honest empirical claim is therefore:

A scaling relation was observed over this specified range.

Not:

The object is infinitely self-similar.

## Fractality Requires More Than Appearance

A branching image can look fractal without satisfying a stable scaling law.

A noisy time series can produce a visually attractive log–log line over a narrow range.

A hierarchy can be nested without being self-similar.

A repeated symbol can be recursive without possessing fractional dimension.

A responsible fractal analysis should declare:

1. the measured object;
2. the scale range;
3. the dimension estimator;

4. uncertainty;
5. alternative distributions;
6. finite-size effects;
7. sensitivity to preprocessing;
8. whether exact or statistical self-similarity is claimed.

## Recursion Is Not Fractality

The Fibonacci sequence is recursive:

$$F_{n+1} = F_n + F_{n-1}.$$

It is not itself a spatial fractal merely because it is recursively generated.

Bayesian updating is recursive:

$$p_{n+1}(\theta) \propto p(D_{n+1} \mid \theta) p_n(\theta).$$

It is not automatically fractal.

Gradient descent is recursive:

$$\theta_{n+1} = \theta_n - \eta \nabla L(\theta_n).$$

It is not automatically self-similar.

The Cathedral Equation is recursive when evaluated through time:

$$C_{n+1} = C_n + \Delta t (G_n - L_n).$$

It is not thereby a fractal equation.

Every fractal construction is recursive in some sense.
Not every recursion is fractal.

## Nested Hierarchy Is Not Necessarily Fractal

The sequence:

galaxy → star system → planet → ecosystem → organism

is an architectural hierarchy.

It is not evidence that:

- a galaxy is a scaled organism;
- a planet is a miniature cell;
- every level possesses the same fractal dimension;
- identical equations govern all levels.

Book I's recursive-differentiation language means that later levels inherit prior organization while introducing new capacities. It does not require exact self-similarity.

## Ratios

For quantities $a$and $b$, with $b \neq 0$, their ratio is:

$$\rho = \frac{a}{b}.$$

If $a$and $b$have the same units, $\rho$is dimensionless.

If their units differ, the ratio carries units.

For two ratios:

$$\frac{a}{b} = \frac{c}{d},$$

the proportion is equivalent to:

$$ad = bc,$$

when $b, d \neq 0$.

### Equation Passport 25.10

*Ratio and Proportion*

$$\rho = \frac{a}{b}\frac{a}{b} = \frac{c}{d} \Longleftrightarrow ad = bc$$

**Status:** [¿]

**Numerator:** $a$

**Denominator:** $b$

**Ratio:** $\rho$

**Dimensionless condition:** $a$ and $b$ possess identical units

**Required declarations:**

- orientation of ratio;
- units;
- denominator choice;
- zero handling;
- measurement uncertainty;
- whether the ratio is instantaneous or averaged.

**What a ratio reveals:** Relative magnitude

**What it does not establish:**

- causation;
- optimality;
- balance;
- fairness;
- universality.

**Architectural significance:** Ratios make relationship measurable without requiring the related quantities to be equal.

## Ratio Orientation Matters

The ratios:

$$\frac{a}{b}$$

and:

$$\frac{b}{a}$$

contain related but differently oriented information.

If:

$$\frac{a}{b}=4,$$

then:

$$\frac{b}{a} = \frac{1}{4}.$$

A pyramid's rise per run is not numerically identical to its run per rise.

A risk–benefit ratio is not the same statement as a benefit–risk ratio.

Every ratio must declare its orientation.

## Composition

For positive or nonnegative component quantities:

$$x_1, \ldots, x_m,$$

define:

$$p_i = \frac{x_i}{\sum_{j=1}^{m} x_j}.$$

Then:

$$p_i \geq 0$$

and:

$$\sum_{i=1}^{m} p_i = 1.$$

The composition vector belongs to the simplex:

$$p \in \Delta^{m-1}.$$

A composition records proportion.

It does not record absolute magnitude.

Two systems may possess the same:

$$p$$

while one is ten times larger in every component.

## The Golden Ratio

Let a line be divided into a larger segment $a$and a smaller segment $b$, with:

$$a>b>0.$$

Extreme and mean ratio requires:

$$\frac{a+b}{a}=\frac{a}{b}=\varphi.$$

Then:

$$\varphi=1+\frac{1}{\varphi}.$$

Multiplying by $\varphi$:

$$\varphi^2=\varphi+1.$$

The positive solution is:

$$\varphi=\frac{1+\sqrt{5}}{2}\approx 1.6180339887.$$

The normalized component shares are:

$$\frac{a}{a+b}=\frac{1}{\varphi}\approx 0.618034,$$

and:

$$\frac{b}{a+b}=\frac{1}{\varphi^2}\approx 0.381966.$$

The relation is explicitly unequal.

Yet it is recursively self-consistent.

Euclid's construction of extreme and mean ratio is preserved in Book VI, Proposition 30.

## Equation Passport 25.11

*Golden-Ratio Recursion*

$$\varphi = 1 + \frac{1}{\varphi} \varphi^2 = \varphi + 1 \varphi = \frac{1+\sqrt{5}}{2}$$

**Status:** $[THM]$

**Domain:** Positive real numbers

**Geometric interpretation:**

$$\frac{\text{whole}}{\text{larger}} = \frac{\text{larger}}{\text{smaller}}$$

**Normalized shares:**

$$\left(\frac{1}{\varphi}, \frac{1}{\varphi^2}\right) \approx (0.618, 0.382)$$

**Recursive significance:** The ratio reproduces itself after removing the larger unit relationship

**What it does not establish:**

- universal biological optimality;
- universal beauty;
- universal architectural superiority;
- cosmic necessity;
- required health ratio;
- proof of SDAE.

**Architectural significance:** A coherent proportion may be unequal while preserving a recursive relational identity.

# Fibonacci Convergence

Let:

$$F_{n+1} = F_n + F_{n-1},$$

with:

$$F_0 = 0, F_1 = 1.$$

Define:

$$r_n = \frac{F_{n+1}}{F_n}$$

for $n \geq 1$.

Then:

$$r_n = 1 + \frac{1}{r_{n-1}}.$$

If the sequence of ratios converges to $r$, then:

$$r = 1 + \frac{1}{r},$$

so:

$$r = \varphi.$$

The golden ratio therefore emerges as the fixed point of the ratio recursion.

This is mathematically beautiful.

It remains one specific recursive architecture.

## The Golden-Ratio Boundary

The golden ratio should be used when:

- the geometric construction requires it;
- a recurrence produces it;
- measured evidence supports it;
- the ratio is relevant to the modeled question.

It should not be inserted merely because a figure looks harmonious.

An attractive approximation to $1.618 \neq$ Evidence of a governing golden-ratio mechanism.

## Dimensionless Control Ratios

Many scientific systems are governed by relative rates or forces rather than absolute values alone.

A generic dimensionless ratio is:

$$\Pi = \frac{\text{process scale A}}{\text{process scale B}}.$$

Examples of interpretation may include:

$$\Pi \ll 1:$$

process B dominates.

$$\Pi \gg 1:$$

process A dominates.

$$\Pi \approx 1:$$

the two declared scales are comparable.

But:

$$\Pi \approx 1$$

does not mean that every component of the system contributes equally.

A ratio becomes informative only after the numerator and denominator are physically or operationally defined.

## Dynamic Balance

Let:

$$x(t)$$

be an accumulated quantity.

Let:

$$G(x,t) \geq 0$$

be its generation or inflow rate.

Let:

$$L(x,t) \geq 0$$

be its loss or outflow rate.

Then:

$$\frac{dx}{dt} = G(x,t) - L(x,t).$$

Point balance occurs when:

$$G(x,t) = L(x,t).$$

Then:

$$\dot{x} = 0.$$

The quantity remains stationary at that moment.

The underlying flows may remain nonzero.

## Equation Passport 25.12

*Dynamic Balance*

$$\dot{x} = G - L$$

**Status:** $[¿/MOD]$

**State quantity:** $x$

**Generation or inflow:** $G$

**Loss or outflow:** $L$

**Instantaneous balance:**

$$G = L$$

**Units:**

$$[G] = [L] = [x] / \text{time}$$

**What balance establishes:** Zero net accumulation in the declared variable

**What it does not establish:**

- inactivity;
- equal component contributions;
- health;
- moral fairness;
- whole-system coherence;
- permanent stability.

**Architectural significance:** Persistence may be produced by continuing unequal activity whose net relationship preserves the state.

## Balance Is Not Fifty–Fifty

Suppose total generation is:

$$G = g_1 + g_2 + g_3$$

with:

$$(g_1, g_2, g_3) = (6,3,1).$$

Then:

$$G = 10.$$

Suppose total loss is:

$$L = l_1 + l_2 + l_3$$

with:

$$(l_1, l_2, l_3) = (4,4,2).$$

Then:

$$L = 10.$$

Therefore:

$$G = L$$

and:

$$\dot{x} = 0.$$

Yet no universal 50/50 distribution appears among the internal processes.

The generation composition is:

$$p_G = (0.6, 0.3, 0.1).$$

The loss composition is:

$$p_L = (0.4, 0.4, 0.2).$$

Balance occurs at the aggregate relation:

$$\sum_i g_i = \sum_j l_j.$$

It does not require:

$$g_i = l_i$$

for every component.

Nor does it require:

$$g_1 = g_2 = g_3.$$

This is the mathematical core of the SDAE statement:

Balance is organized proportion, not universal equality.

# Vector Flux Balance

Let:

$$x(t) \in R^m$$

be a vector of accumulated quantities.

Let:

$$v(t) \in R^q$$

be process or flux rates.

Let:

$$S \in R^{m \times q}$$

be a structure matrix describing how each process changes each quantity.

Then:

$$\dot{x} = S v .$$

A steady-state flux condition is:

$$S v^{¿} = 0.$$

This does not require:

$$v^{¿} = 0.$$

Nonzero internal flows may lie in the null space of $S$.

## Equation Passport 25.13

*Vector Flux Balance*

$$\dot{x} = S v S v^{¿} = 0$$

**Status:** $[¿ / MOD]$

**Accumulated state:** $x$

**Flux vector:** $v$

**Structure matrix:** $S$

**Steady-state condition:**

$$\dot{x} = 0$$

**Permitted condition:**

$$v^{\text{¿}} \neq 0$$

**What it reveals:** Macroscopic constancy can coexist with active internal circulation

**Failure conditions:**

- omitted fluxes;
- wrong system boundary;
- changing storage;
- delayed processes;
- measurement inconsistency.

**Architectural significance:** The Cathedral can remain standing while energy, matter, information, responsibility, and repair continually move through it.

## Balance Manifolds

For dynamics:

$$\dot{x} = F(x),$$

the balance set is:

$$B = \{x : F(x) = 0\}.$$

The balance condition may define:

- one point;
- several points;
- a curve;
- a surface;
- a higher-dimensional manifold.

Balance is therefore not always one ideal state.

A system may possess several viable balanced organizations.

This is important for:

- biological variability;
- cultural plurality;
- alternative institutional forms;
- multiple strategies of adaptation.

One coherence requirement may admit many balanced realizations.

## Point, Cycle, Bounded, and Adaptive Balance

### Point Balance

$$G(t)=L(t)$$

at each relevant moment.

### Cycle Balance

For a cycle of duration $T$:

$$x(t+T)=x(t).$$

Therefore:

$$\int_{t}^{t+T}[G(s)-L(s)]ds=0.$$

Generation and loss need not be equal at every instant.

### Bounded Balance

A variable remains within an admissible range:

$$x_{min}\leq x(t)\leq x_{max}.$$

It may fluctuate continually.

### Adaptive Balance

The viable reference changes:

$$x^{¿}=x^{¿}(t,c),$$

where $c$ is context.

A winter physiological pattern need not equal a summer pattern.

A child's viable organization need not equal an elder's.

A recovering system need not occupy its pre-disturbance state.

## Equation Passport 25.14

***Cycle-Averaged Balance***

$$x(t+T)-x(t)=\int_{t}^{t+T}(G-L)\,ds$$

**Status:** $[THM/MOD]$

**Cycle balance:**

$$x(t+T)=x(t)$$

implies:

$$\int_{t}^{t+T}(G-L)\,ds=0.$$

**What it permits:**

- oscillation;
- phase difference;
- temporary accumulation;
- temporary depletion;
- rhythmic exchange.

**What it does not establish:**

- zero instantaneous error;
- equal amplitudes;
- absence of cost;
- resilience outside the cycle.

**Architectural significance:** Coherence may be preserved through a complete rhythm rather than at every isolated moment.

## Dynamic Balance and Stability

Balance and stability must be distinguished.

Suppose:

$$\dot{x} = F(x)$$

and:

$$F(x^{*}) = 0.$$

The state $x^{*}$ is balanced.

It is locally stable only if nearby states remain near or return under the relevant stability definition.

A balanced state may be unstable.

For example:

$$\dot{x} = kx, k > 0,$$

has equilibrium:

$$x^{*} = 0.$$

Yet any nonzero perturbation grows:

$$x(t) = x_0 e^{kt}.$$

Thus:

$$\dot{x} = 0$$

at $x = 0$, but the equilibrium is unstable.

$$\text{Balance Condition} \neq \text{Stability of Balance}.$$

## Target Ratios

Suppose two positive processes should remain in target proportion:

$$\rho^{¿}=\frac{a^{¿}}{b^{¿}}.$$

The observed ratio is:

$$\rho(t)=\frac{a(t)}{b(t)}.$$

A scale-symmetric ratio error is:

$$e_{\rho}(t)=\ln\left[\frac{\rho(t)}{\rho^{¿}(t)}\right].$$

Then:

$$\rho=2\rho^{¿}$$

gives:

$$e_{\rho}=\ln 2,$$

while:

$$\rho=\frac{1}{2}\rho^{¿}$$

gives:

$$e_{\rho}=-\ln 2.$$

A proportional correction model is:

$$\dot{e}_{\rho}=-k\,e_{\rho}+d(t),k>0.$$

## Equation Passport 25.15

***Ratio-Control Error***

$$e_{\rho}=\ln\left(\frac{\rho}{\rho^{\text{¿}}}\right)\dot{e}_{\rho}=-k\,e_{\rho}+d\left(t\right)$$

**Status:** $\left[MOD\right]$

**Observed ratio:** $\rho$

**Target ratio:** $\rho^{\text{¿}}$

**Correction rate:** $k$

**Disturbance:** $d\left(t\right)$

**Balanced ratio:**

$$e_{\rho}=0.$$

**Advantages of log-ratio error:**

- dimensionless;
- multiplicatively symmetric;
- distinguishes proportional deviation.

**Required declarations:**

- why the target ratio is appropriate;
- whether the target changes;
- positivity of quantities;
- measurement error;
- allowable range.

**What it does not establish:** That one universal target ratio governs every state or person

**Architectural significance:** Dynamic balance means correcting toward the right relation, not forcing equal magnitudes.

# The Target Ratio May Change

Let:

$$\rho^{\text{¿}}=\rho^{\text{¿}}\left(x,t,c,h\right),$$

where:

- $x$ is current state;
- $t$ is time;
- $c$ is context;
- $h$ is history.

A healing system may require a different ratio of:

- activity to rest;
- inflammation to repair;
- openness to protection;
- conservation to exploration;

than a fully recovered system.

The governing question is not:

Is the ratio equal?

It is:

**Is the ratio appropriate to the system's present state, purpose, and future viability?**

## Recursive Differentiation

Book I's architectural recognition can now receive a formal descendant.

Let:

$$R_n$$

be organized relational state at stage $n$.

Let:

$$H_n$$

be an inheritance map preserving selected prior organization.

Let:

$$D_n$$

be a differentiation operator.

Let:

$$U_n$$

be developmental, environmental, or contextual input.

Let:

$$\eta_n$$

represent unresolved novelty, variation, or perturbation.

Then:

$$R_{n+1} = D_n[H_n,(R_n), U_n \eta_n].[ARC/MOD]$$

The later state is neither created from nothing nor copied unchanged.

It is inherited and differentiated.

## Equation Passport 25.16

*Recursive Differentiation*

$$R_{n+1} = D_n[H_n,(R_n), U_n \eta_n]$$

**Status:** $[ARC/MOD/CONJ]$

**Relational state:** $R_n$

**Inheritance operator:** $H_n$

**Differentiation operator:** $D_n$

**Input:** $U_n$

**Variation or novelty:** $\eta_n$

**Required declarations:**

- mathematical type of $R_n$;

- preserved invariants;
- differentiation rule;
- state and scale interpretation;
- stochastic structure;
- observables.

**Possible mathematical types:**

- graph;
- hypergraph;
- tensor;
- manifold;
- category;
- probability distribution;
- hybrid system.

**What it provides:** A formal architecture of inherited transformation

**What it does not establish:**

- one universal mechanism of differentiation;
- inevitable progress;
- exact fractality;
- predetermined emergence;
- conscious purpose.

**Architectural significance:** Every differentiated stage carries the past forward while becoming more than a copy of the past.

## Branching Differentiation

A state may generate several descendants:

$$R_{n+1,k}=D_{n,k}\left(R_n,U_{n,k},\eta_{n,k}\right),k=1,\ldots,m_n.$$

This produces a branching tree or graph.

The descendants share ancestry.

They need not share destiny.

A branching architecture can represent:

- cell lineages;
- evolutionary divergence;
- conceptual specialization;
- institutional differentiation;
- alternative policy futures.

Again:

$$\text{Common Origin} \neq \text{Identical Present Mechanism}.$$

# Invariants and Novel Capacities

Let:

$$I_j(R)$$

be a declared invariant or approximately preserved feature.

Inheritance may require:

$$I_j(R_{n+1}) \approx I_j(R_n)$$

for selected $j$.

Let:

$$C(R_n)$$

be the capability set available at stage $n$.

Novel emergence occurs when:

$$C(R_{n+1}) \not\subseteq T_n[C(R_n)]$$

for the allowed transformations $T_n$.

This means the later system possesses at least one capability not obtained by merely relabeling an earlier capability.

## Equation Passport 25.17

*Inheritance and Novelty*

$$I_j(R_{n+1}) \approx I_j(R_n)$$

while:

$$C\left(R_{n+1}\right) \not\subseteq T_n\left[C\left(R_n\right)\right].$$

**Status:** $\left[ARC/MOD/CONJ\right]$

**Inherited invariants:** $I_j$

**Capability set:** $C\left(R\right)$

**Permitted relabeling or transfer:** $T_n$

**Interpretation:** Selected organization persists while new capability appears

**Required work:**

- define capability operationally;
- distinguish latent prior capability from genuine novelty;
- test alternative decompositions;
- identify scale dependence.

**What it does not establish:** Metaphysical emergence merely because the model changes vocabulary

**Architectural significance:** Continuity and novelty need not be opposites.

## The Inheritance–Novelty Tension

Too much inheritance without differentiation yields:

- rigidity;
- repetition;
- stagnation;
- inability to answer changed conditions.

Too much differentiation without inheritance yields:

- fragmentation;
- loss of identity;
- inability to transmit learning;
- collapse of continuity.

A symbolic pair of normalized quantities may be:

$$h_n \in [0,1]$$

for inherited-structure preservation, and:

$$v_n \in [0,1]$$

for novelty generation.

Coherence does not require:

$$h_n = v_n = 0.5.$$

The appropriate relation depends upon:

- disturbance;
- developmental stage;
- system reserve;
- risk;
- purpose.

A stable environment may support greater inheritance.

A rapidly changing environment may require greater novelty.

A threatened identity may require strong boundary preservation.

A rigid institution may require deliberate differentiation.

## Recursive Dual-Aspect Dynamics

Chapter Twenty-Four defined:

$$\zeta_n = (x_{E,n}, x_{I,n}, b_n).$$

A recursive dual-aspect update is:

$$x_{E,n+1} = F_E(x_{E,n}, x_{I,n}, b_n, u_{E,n}, d_{E,n}),$$
$$x_{I,n+1} = F_I(x_{I,n}, x_{E,n}, b_n, u_{I,n}, d_{I,n}), b_{n+1} = F_B(b_n, x_{E,n}, x_{I,n}, c_n).$$

Compactly:

$$\zeta_{n+1} = F(\zeta_n, u_n, d_n, c_n).$$

## Equation Passport 25.18

*Recursive Dual-Aspect Update*

$$\zeta_{n+1} = F(\zeta_n, u_n, d_n, c_n)$$

**Status:** $[ARC/MOD/CONJ]$

**Joint state:**

$$\zeta_n = (x_{E,n}, x_{I,n}, b_n)$$

**Exterior state:** $x_{E,n}$

**Interior modeled state:** $x_{I,n}$

**Interface:** $b_n$

**Context:** $c_n$

**Possible recursive processes:**

- perception;
- attention;
- action;
- feedback;
- learning;
- memory;
- reappraisal;
- narrative revision.

**What it does not establish:**

- independent immaterial causation;
- universal state coordinates;
- complete observability;
- one timescale;
- physical status of symbolic variables.

**Architectural significance:** One participating life becomes historical because each exterior and interior state enters the conditions of the next.

## Recursive Learning as an Example

Suppose a model parameter is:

$$\theta_n.$$

Prediction error is:

$$e_n = y_n - \hat{y}_n(\theta_n).$$

A learning update is:

$$\theta_{n+1} = \theta_n + \eta_n \Psi(e_n, x_n).$$

The update changes future prediction.

Experience therefore does not merely pass through the system.

It changes the rule through which later experience will be interpreted.

This is recursion at a second order:

$$\text{State changes}$$

and:

$$\text{the state-update architecture changes}.$$

## Recursive Self-Modeling

Let:

$$m_n$$

be a system's model of itself.

Let:

$$y_n$$

be new evidence.

Then:

$$m_{n+1} = U(m_n, y_n).$$

Action depends upon the self-model:

$$a_n = \pi(m_n, x_n).$$

Action changes the world:

$$x_{n+1} = F(x_n, a_n, d_n).$$

The resulting consequence becomes new evidence:

$$y_{n+1} = H(x_{n+1}, a_n).$$

Thus:

$$\text{Self-Model} \rightarrow \text{Action} \rightarrow \text{Consequence} \rightarrow \text{Revised Self-Model}.$$

This recursive loop will become central in Chapter Twenty-Seven's Bridge Species architecture.

## Scale-Indexed Dual-Aspect States

Let:

$$\zeta_n^{(l)} = \left(x_{E,n}^{(l)}, x_{I,n}^{(l)}, b_n^{(l)}\right)$$

be the joint state at time step $n$and descriptive scale $l$.

Temporal evolution is:

$$\zeta_{n+1}^{(l)} = F_l\left(\zeta_n^{(l)}\right).$$

Scale transformation is:

$$\zeta_n^{(l+1)} = S_l\left(\zeta_n^{(l)}\right).$$

Two procedures are now possible.

### Evolve, Then Coarse-Grain

$$S_l\left[F_l\left(\zeta_n^{(l)}\right)\right].$$

### Coarse-Grain, Then Evolve

$$F_{l+1}\left[S_l\left(\zeta_n^{(l)}\right)\right].$$

They need not agree.

## Scale–Dynamics Commutation Defect

Define:

$$\Delta_{l,n}=d_{l+1}\left(S_l,\left[F_l\left(\zeta_n^{(l)}\right)\right]F_{l+1}\left[S_l\left(\zeta_n^{(l)}\right)\right]\right).$$

If:

$$\Delta_{l,n}=0,$$

the scale map and evolution commute exactly for that state.

If:

$$\Delta_{l,n}\approx 0,$$

the coarse model approximately preserves the fine dynamics.

If:

$$\Delta_{l,n}$$

is large, the coarse-level model fails to reproduce the coarse-grained fine evolution.

### Equation Passport 25.19

*Scale–Dynamics Commutation Defect*

$$\Delta_{l,n}=d\left(S_l,\circ F_l\left(\zeta_n^{(l)}\right)F_{l+1}\circ S_l\left(\zeta_n^{(l)}\right)\right)$$

**Status:** $[ARC/MOD/TEST]$

**Fine-scale evolution:** $F_l$

**Coarse-scale evolution:** $F_{l+1}$

**Scale map:** $S_l$

**State-space distance:** $d$

**Exact scale consistency:**

$$\Delta_{l,n}=0.$$

**Approximate scale consistency:**

$$\Delta_{l,n}\le\varepsilon.$$

**What a large defect may indicate:**

- omitted coarse variable;
- memory dependence;
- nonlinear aggregation;
- emergent constraint;
- wrong boundary;
- inadequate scale map;
- genuinely different effective mechanism.

**Architectural significance:** “The mechanism changes” can become a measurable failure of scale-level interchangeability.

## Emergence as Failure of Naive Scale Commutation

Suppose the fine-level variables evolve correctly.

Suppose simple averaging produces a coarse state.

If:

$$\Delta_{l,n}\gg 0,$$

the macroscopic future cannot be predicted merely by applying the assumed coarse law to the averaged present.

A better coarse model may require:

- interaction terms;
- network topology;
- memory;
- boundary effects;
- collective variables;
- new effective parameters.

The formalism need not proclaim mysterious emergence whenever a simple scale map fails.

It must first ask whether the coarse model omitted necessary structure.

## Multiscale Dynamic Balance

Let each scale possess a coherence state:

$$C^{(l)}(t).$$

Its dynamics are:

$$\dot{C}^{(l)} = G^{(l)} - L^{(l)} + J_{l-1\to l} - J_{l\to l+1},$$

where:

- $J_{l-1\to l}$ is support entering from a finer or nested scale;
- $J_{l\to l+1}$ is support or cost exported to another scale.

A local scale may appear balanced:

$$G^{(l)} = L^{(l)}$$

while depending upon unsustainable inflow:

$$J_{l-1\to l} > 0.$$

Or it may preserve itself by exporting loss:

$$J_{l\to l+1} \gg 0.$$

## Equation Passport 25.20

*Multiscale Coherence Balance*

$$\dot{C}^{(l)} = G^{(l)} - L^{(l)} + J_{l-1 \to l} - J_{l \to l+1}$$

**Status:** $[ARC/MOD]$

**Scale coherence:** $C^{(l)}$

**Local generation:** $G^{(l)}$

**Local loss:** $L^{(l)}$

**Cross-scale transfer:** $J$

**Required declarations:**

- scale hierarchy;
- boundary;
- transfer units;
- direction;
- external support;
- exported harm;
- time horizon.

**What it prevents:** Declaring one level balanced while concealing the depletion of another

**Architectural significance:** Right relationship must hold not only within a scale, but among the scales that support one another.

# Scale Balance Is Not Scale Equality

Cells, organs, organisms, families, institutions, and civilizations do not possess equal:

- size;
- energy use;
- authority;
- timescale;
- function;
- information capacity.

Multiscale coherence does not require:

$$C^{(1)} = C^{(2)} = \cdots = C^{(L)}.$$

It requires that each scale remain within its viable range and that cross-scale relations do not progressively destroy another indispensable level.

## Exterior Recursion

Exterior systems may exhibit recursion through:

- repeated physical update;
- biological development;
- feedback control;
- network propagation;
- ecological succession;
- institutional policy cycles.

A general exterior recursion is:

$$x_{E,n+1} = F_E(x_{E,n}, u_n, d_n).$$

Its empirical observables may include:

- trajectory;
- fixed point;
- period;
- Lyapunov exponent;
- scaling exponent;
- recovery time;
- state distribution.

## Interior Recursion

Interior life may exhibit modeled recursion through:

- expectation affecting perception;
- attention affecting memory;
- memory affecting interpretation;
- interpretation affecting future attention;
- narrative affecting choice;
- consequence revising narrative.

A symbolic model is:

$$x_{I,n+1} = F_I(x_{I,n}, e_n, m_n, v_n, r_n)[ARC/MOD],$$

where:

- $e_n$ is experience;
- $m_n$ is active memory;
- $v_n$ is value;
- $r_n$ is reflection or relationship.

Interior recursion does not imply that a thought is a fractal object.

It means earlier organization changes the conditions through which later experience is lived.

## Narrative Recursion

Let:

$$N_n$$

be the current life narrative.

An event occurs:

$$e_{n+1}.$$

The narrative is revised:

$$N_{n+1} = R(N_n, e_{n+1}, m_n, c_n).$$

But the current narrative also affects how the event is interpreted:

$$\hat{e}_{n+1} = I(e_{n+1}, N_n).$$

Thus:

$$N_n \to \hat{e}_{n+1} \to N_{n+1}.$$

The narrative interprets the event.

The interpreted event revises the narrative.

A rigid narrative minimizes revision.

A coherent narrative preserves identity while allowing evidence to reorganize meaning.

## Recursive Social Organization

Institutions also recurse.

Let policy be:

$$P_n.$$

Policy produces consequences:

$$Y_n = F(P_n, E_n).$$

Consequences are measured:

$$D_n = H(Y_n).$$

The institution updates:

$$P_{n+1} = U(P_n, D_n, V_n).$$

An institution loses recursive coherence when:

- consequences are hidden;
- measurement is distorted;
- criticism is punished;
- policy cannot be revised;
- inherited purpose becomes immune from reality.

## The Holofractal Boundary

SDAE may use **holofractal** architecturally to recognize recurring part–whole and scale relationships.

The term must remain status-marked.

A system may be called holofractal in an architectural sense when:

- local organization carries partial information about larger organization;
- motifs recur across levels;
- differentiated parts participate in a larger whole;
- reconstruction remains possible after partial loss;
- recursive scale maps preserve selected invariants.

This does not establish:

- exact self-similarity;
- one fractal dimension across all levels;
- literal holographic encoding;
- the whole in complete resolution within every part;
- a physical holographic universe;
- a quantum mechanism of consciousness.

Chapter Twenty-Six will examine the distributed part–whole claim directly.

## The Spirit Field Boundary

The Spirit Field remains:

$$S_f[SYM/ARC]$$

as symbolic language for organized interior coherence.

A symbolic recursive relation may be written:

$$S_{f,n+1} = R_{sym}(S_{f,n}, E_n, M_n, V_n, P_n, C_n),$$

where:

- $E_n$ is lived experience;
- $M_n$ is memory;
- $V_n$ is value;
- $P_n$ is purpose;
- $C_n$ is compassion or context.

This may represent the reflective recognition that:

- gratitude may deepen through remembered inheritance;
- hope may be revised by evidence;

- forgiveness may change the future influence of past injury;
- purpose may become more responsible through consequence;
- compassion may widen the boundary of participation.

It does not establish:

- a physical recursive field;
- a measurable quantum variable;
- a new force;
- a clinical metric;
- literal fractality of interior life.

# A Computational Preview

Chapter Twenty-Eight will provide the full computational program.

This chapter can already identify several experiments.

## Experiment One — Logistic Bifurcation

For:

$$x_{n+1} = r x_n (1 - x_n),$$

vary:

$$r \in [0,4].$$

For each $r$:

1. choose $x_0$;
2. discard a transient;
3. record the later orbit;
4. estimate periodicity;
5. estimate the Lyapunov exponent;
6. plot the attractor against $r$.

**Test:** Identify transitions from fixed point to periodicity and chaos.

**Failure condition:** Treating finite numerical output as proof of every asymptotic property.

## Experiment Two — Iterated-Function-System Reconstruction

Choose contractions:

$$w_1, \dots, w_N.$$

Iterate:

$$A_{n+1} = ¿\, i\, w_i \left( A_n \right).$$

**Test:** Compare measured box-counting dimension with the similarity dimension.

**Failure condition:** Separation assumptions fail or finite resolution biases the estimate.

## Experiment Three — Unequal Dynamic Balance

Use:

$$\dot{x} = \sum_i g_i - \sum_j l_j.$$

Choose unequal component rates with equal totals.

Perturb one contribution.

**Test:** Determine which contributions most strongly alter the balance margin.

**Prediction:** Influence depends upon sensitivity and network placement, not merely whether the component's share is close to fifty percent.

## Experiment Four — Scale Commutation

Construct:

$$F_l, F_{l+1}, S_l.$$

Compute:

$$\Delta_{l,n} = d\left( S_l, F_l \left( \zeta \right) F_{l+1} S_l \left( \zeta \right) \right).$$

**Test:** Determine where the coarse model ceases to reproduce coarse-grained fine dynamics.

**Interpretation:** A large defect identifies missing effective structure—not automatically metaphysical emergence.

### Experiment Five — Recursive Dual-Aspect Prediction

Compare:

$$M_0 : x_{E,n+1} = F_E\left(x_{E,n}\right)$$

with:

$$M_1 : x_{E,n+1} = F_E\left(x_{E,n}, x_{I,n}, b_n\right).$$

**Test:** Determine whether interior-state estimates add reproducible out-of-sample predictive information beyond exterior history and context.

**Failure condition:** Reports leak the target, share measurement bias, or merely proxy an omitted exterior variable.

## Testable Expectations and Open Problems

### Test 25.1 — Recursive-State Dependence

When a system is genuinely recursive, prediction from the present state should outperform prediction from an unrelated state after controlling for obvious covariates.

### Test 25.2 — Fixed-Point Stability

Perturbations near an attracting fixed point should contract at a rate consistent with the local derivative or Jacobian under the model.

### Test 25.3 — Periodic Recurrence

A claimed period-$p$ orbit should satisfy:

$$d\left(F^p, \left(x\right) x\right) \le \varepsilon$$

while failing the corresponding condition for smaller periods.

### Test 25.4 — Positive Lyapunov Growth

A claim of deterministic chaos should demonstrate reproducible positive trajectory divergence under a valid reconstruction and should distinguish sensitivity from measurement noise.

### Test 25.5 — Fractal Scaling

A claimed fractal dimension should remain approximately stable across a declared scaling interval and under reasonable changes of estimator.

### Test 25.6 — Exact Versus Statistical Self-Similarity

The evidence should distinguish exact geometric reconstruction from equality of statistical distributions.

### Test 25.7 — Mechanism Difference

Cross-scale structural similarity should survive explicit comparison even when domain-specific mechanism models remain demonstrably different.

### Test 25.8 — Unequal Balance

A balanced system should permit nonuniform contribution vectors:

$$p \neq \left(\frac{1}{m}, \dots, \frac{1}{m}\right)$$

while preserving its declared viability condition.

### Test 25.9 — Context-Dependent Target Ratio

A fixed-ratio model and an adaptive-ratio model should be compared out of sample.

### Test 25.10 — Scale Commutation

The commutation defect should predict where a coarse model loses reliability.

## Open Problem 25.1

Which feature spaces best distinguish genuine structural recurrence from superficial visual analogy?

## Open Problem 25.2

Can a general taxonomy classify cross-scale relations as:

- exact;
- statistical;
- topological;
- causal;
- functional;
- architectural;
- symbolic?

## Open Problem 25.3

How should recursive differentiation be operationalized when later levels possess state variables unavailable at earlier levels?

## Open Problem 25.4

Can emergence be quantified through predictive gain obtained by new effective variables after coarse-graining?

## Open Problem 25.5

How should interior recursive states be compared across persons without assuming one universal experiential coordinate system?

## Open Problem 25.6

Can dynamic-balance models preserve noncompensable dignity constraints rather than reducing every relation to one scalar target?

## Common Failures of Recursive and Fractal Architecture

### Recursion–Repetition Collapse

A transformed later state is treated as an unchanged copy.

### Recursion–Progress Collapse

Every recursive update is assumed to improve the system.

### Recursion–Fractal Collapse

Any recurrence relation is called fractal.

### Fractal-by-Appearance

Visual complexity is treated as proof of fractal dimension.

### Infinite-Scaling Assumption

A finite empirical scaling range is described as infinite self-similarity.

### Dimension-Estimator Absolutism

One numerical estimate is treated as the object's unique true complexity.

### Exact–Statistical Collapse

Approximate distributional similarity is treated as exact geometric identity.

### Self-Similarity–Whole-in-Part Collapse

A scaled resemblance is treated as complete encoding of the whole in every part.

### Nested-Hierarchy Fractalization

Every level of a hierarchy is treated as a scaled copy of every other level.

### Mechanism Identity

Similar shapes or equations are treated as evidence of identical causal production.

### Scale-Invariance Export

A scaling law valid near one critical regime is extended to all states and scales.

### Renormalization Universalism

A domain-specific physical renormalization method is applied metaphorically as though it were an empirical law of consciousness or civilization.

### Universality–Ontology Collapse

Shared mathematical exponents are treated as proof that systems are fundamentally the same kind of thing.

### Chaos–Randomness Collapse

Deterministic sensitivity is described as random generation.

### Chaos Romanticism

Unpredictable dynamics are automatically called creative, healthy, or spiritually significant.

### Bifurcation–Moral Transformation Collapse

A mathematical change of attractor is treated as moral or spiritual rebirth without evidence.

### Fixed-Point Worship

A stable state is treated as desirable merely because it reproduces itself.

### Periodicity–Health Collapse

Regular recurrence is treated as proof of whole-system health.

## Golden-Ratio Universalism

$\varphi$ is inserted wherever an unequal proportion appears.

## Approximation Hunting

Measurements are rounded until they resemble $1.618$.

## Beauty–Mechanism Collapse

A pleasing form is treated as evidence that the golden ratio generated it.

## Ratio Orientation Erasure

$\frac{a}{b}$ and $\frac{b}{a}$ are treated as interchangeable.

## Ratio–Causation Collapse

A stable quotient is treated as proof that one quantity controls the other.

## Composition–Magnitude Collapse

Equal proportions are treated as equal total quantity.

## Balance–Equality Collapse

Dynamic balance is treated as equal contribution by all components.

## Balance–Fifty–Fifty Collapse

A two-sided metaphor is mistaken for an arithmetic requirement.

## Steady-State–Inactivity Collapse

Zero net accumulation is treated as absence of internal process.

## Balance–Stability Collapse

A state satisfying the balance equation is assumed to be stable.

## Stability–Goodness Collapse

A stable but harmful organization is treated as coherent.

## Point-Balance Absolutism

A rhythmically balanced system is judged only at one instant.

## Cycle-Balance Erasure

Temporary deviations are called failures despite zero net change over the viable cycle.

## Target-Ratio Fixation

A context-dependent relationship is forced toward one permanent number.

## Adaptive-Balance Relativism

Every observed ratio is treated as acceptable merely because context varies.

## Aggregate-Balance Concealment

A balanced total hides one collapsing component.

## Boundary Manipulation

External support or exported loss is omitted from the balance equation.

## Scale-Transfer Erasure

Local balance is praised while another scale supplies unsustainable support.

## Inheritance Worship

Prior organization is preserved despite changed reality.

## Novelty Worship

Inherited organization is discarded merely because something new is possible.

## Recursive-History Erasure

Transformation order is ignored.

### Path-Independence Assumption

The same components are assumed to produce the same state regardless of sequence.

### Latent-Fractal Physicalization

A fitted fractal exponent is treated as a new physical field.

### Interior-Fractal Reification

Metaphorical recursive experience is described as measured fractal consciousness without evidence.

### Spirit-Field Fractalization

The Spirit Field is converted into a physical fractal or scale field without empirical support.

## The Recursion, Fractal, Ratio, and Balance Constitution

Before any claim enters this laboratory, it must answer:

1. What object is recurring?
2. What state space contains it?
3. What is the initial state?
4. What is the update rule?
5. Is the rule constant or time-dependent?
6. Is the system deterministic or stochastic?
7. What inputs enter?
8. What disturbances enter?
9. What history is retained?
10. What history is discarded?
11. Is the process iterative?
12. Is it recursive?
13. Is it path-dependent?
14. Is it memoryless?
15. Is recursion being confused with repetition?
16. Is the later state identical to the earlier state?
17. Is it approximately recurrent?
18. What metric defines recurrence?
19. What tolerance is used?
20. What time or depth interval applies?
21. Does a fixed point exist?

22. Is it unique?
23. Is it locally stable?
24. Is it globally stable?
25. What derivative or Jacobian determines stability?
26. Does a periodic orbit exist?
27. What is its minimum period?
28. Is the observed cycle transient?
29. Is a bifurcation claimed?
30. Which parameter changes?
31. What qualitative structure changes?
32. Is the bifurcation local or global?
33. Is chaos claimed?
34. What definition of chaos is used?
35. Is sensitive dependence demonstrated?
36. Is the Lyapunov exponent positive?
37. Is the estimate robust to noise?
38. Is boundedness demonstrated?
39. Is aperiodicity demonstrated?
40. Is deterministic chaos being confused with randomness?
41. Is a universality claim made?
42. What mathematical class supports it?
43. Which quantities are universal?
44. Which quantities remain model-specific?
45. Is universality being confused with mechanism identity?
46. Is a scale transformation used?
47. What does the scale index mean?
48. Is the scale spatial, temporal, organizational, or descriptive?
49. What is coarse-grained?
50. What information is lost?
51. Is the scale map invertible?
52. Is a renormalization group used?
53. What parameter space does it act upon?
54. What fixed point is identified?
55. What directions are relevant or irrelevant?
56. Is a physical renormalization method being exported outside its domain?
57. Is self-similarity claimed?
58. Is it exact?
59. Is it statistical?
60. Is it affine?
61. Is it topological rather than metric?
62. What transformation relates the scales?
63. What error remains after transformation?
64. Is a fractal claimed?
65. What definition is used?
66. What dimension is estimated?
67. Is it similarity dimension?
68. Is it box-counting dimension?
69. Is it Hausdorff dimension?
70. Are these dimensions being treated as interchangeable?
71. What scale range is measured?
72. How many decades of scaling exist?

73. What finite-size effects occur?
74. What measurement resolution applies?
75. What uncertainty accompanies the dimension?
76. Does the result survive another estimator?
77. Does an alternative nonfractal model fit equally well?
78. Is visual resemblance being treated as evidence?
79. Is a hierarchy being called fractal without scaling analysis?
80. Is part–whole encoding being confused with self-similarity?
81. What pattern recurs?
82. What feature map represents it?
83. What mechanism generates it?
84. How is mechanism difference measured?
85. Is the Project Cathedral rule preserved?
86. Is structural analogy clearly distinguished from causal identity?
87. What ratio is used?
88. What is the numerator?
89. What is the denominator?
90. What units remain?
91. Is the ratio dimensionless?
92. Is its orientation declared?
93. Can the denominator approach zero?
94. Is a composition used?
95. Are absolute magnitudes also reported?
96. Is a target ratio claimed?
97. What evidence supports the target?
98. Is the target fixed or adaptive?
99. Is the golden ratio invoked?
100. Does the defining relation actually produce $\varphi$?
101. Is the ratio measured precisely enough?
102. Is the use geometric, recursive, empirical, or symbolic?
103. Is $\varphi$ being treated as universally optimal?
104. Is balance claimed?
105. What state variable is balanced?
106. What generation processes are included?
107. What loss processes are included?
108. Are all rates expressed in compatible units?
109. Is the balance instantaneous?
110. Is it cycle-averaged?
111. Is it bounded-range balance?
112. Is it adaptive balance?
113. Is balance being confused with equal shares?
114. Is fifty–fifty equality being imposed?
115. Can unequal contributions preserve the balance?
116. Are component shares visible?
117. Does aggregate balance conceal local collapse?
118. Does internal flow remain nonzero?
119. Is a flux matrix used?
120. What null-space flows exist?
121. Is the balanced state stable?
122. What perturbations threaten it?

123. Does the balance possess margin?
124. Is the target ratio state-dependent?
125. What correction law applies?
126. What is the cost of correction?
127. Does correction deplete reserve?
128. Is another scale supplying hidden support?
129. Is another scale absorbing exported loss?
130. What is inherited through recursion?
131. What invariants are preserved?
132. What is newly differentiated?
133. What new capability appears?
134. Could the capability have been latent earlier?
135. Is novelty being inferred only from changed vocabulary?
136. Is branching represented?
137. Are descendants treated as identical because they share ancestry?
138. Is path dependence represented?
139. Does transformation order matter?
140. Are exterior states recursive?
141. Are interior states recursively modeled?
142. What report or observation constrains interior recursion?
143. Is narrative recursion distinguished from historical fact?
144. Is self-modeling represented?
145. Can the update rule itself change?
146. Is temporal recursion distinguished from scale recursion?
147. Are time and scale indices being confused?
148. Does evolution commute with coarse-graining?
149. What is the commutation defect?
150. Does a large defect indicate a missing variable?
151. Could a better scale map reduce it?
152. Is emergence being declared prematurely?
153. Is multiscale balance modeled?
154. What transfers occur among scales?
155. Does local viability damage global viability?
156. Does global stability suppress necessary local differentiation?
157. Is holofractal language invoked?
158. Is its architectural status explicit?
159. Is exact mathematical fractality being claimed?
160. Is holographic encoding being claimed?
161. What evidence supports either?
162. Is the Spirit Field invoked?
163. Is its symbolic status explicit?
164. Is any physical fractal field being proposed?
165. What evidence would be required?
166. What computational experiment tests the claim?
167. What null model is used?
168. What parameter range is tested?
169. What sensitivity analysis is performed?
170. What observation would falsify the claim?
171. What remains scale-dependent?
172. What remains mechanism-specific?
173. What remains merely analogous?

174. What remains symbolic?
175. What does the claim not establish?

A claim unable to answer these questions may remain useful as:

$$[¿], [MOD], [ARC], [CONJ], \text{or } [SYM].$$

Its status must remain visible.

## The Architect's Reflection

For many years, I believed recurrence meant that reality was repeating itself.

The same forms.

The same struggles.

The same lessons.

Reality gradually taught another lesson.

Nothing returns unchanged.

A child may inherit a parent's face without inheriting the parent's life.

A student may inherit a teacher's equation without inheriting the question that first gave birth to it.

A civilization may inherit a monument while forgetting the discipline that made the monument possible.

A wound may return through memory, yet the person remembering it is no longer the person who first received it.

Perhaps recursion is not reality circling helplessly around the same point.

Perhaps recursion is how history remembers enough of itself to become capable of another step.

The seed returns in the tree.

But not as another seed alone.

The ancestor returns in the child.

But not as an identical life.

Book I returns in Book II.

But not as repetition.

Its constitutional recognitions now return with:

- variables;
- state spaces;
- thresholds;
- scale maps;
- uncertainties;
- computational tests.

The pattern recurs.

The mechanism changes.

The inheritance remains.

The capacity grows.

Balance likewise became clearer.

I once understood balance as equal halves.

Equal time.

Equal force.

Equal contribution.

Reality showed otherwise.

The heart and kidney do not contribute equally to the same task.

The teacher and student do not carry equal responsibility at every moment.

The parent may carry nearly everything while the infant carries almost nothing.

A recovering body may require far more rest than exertion.

A wounded relationship may require far more listening than speaking.

Equality remains sacred in dignity.

It is not always the geometry of function.

Perhaps the deepest balance is not numerical sameness.

Perhaps it is every differentiated part receiving and carrying the measure appropriate to the life of the whole.

The golden ratio is beautiful not because all balance must become 1.618.

It is beautiful because it demonstrates something more humble:

An unequal relationship can remain internally consistent.

The Cathedral does not stand because every stone bears the same weight.

It stands because each stone bears the right relationship to the load passing through it.

## Quiet Reflection

Think of one pattern that has recurred within your life.

A relationship.

A fear.

A responsibility.

A question.

A hope.

Ask:

Did the event repeat exactly?

Or did an earlier organization shape the conditions of a later one?

What was inherited?

What changed?

What became possible this time that was not possible before?

Now consider one area of your life you call unbalanced.

Do you mean:

- the two sides are unequal;
- the present ratio is inappropriate;
- one process is overloaded;
- one contribution is absent;
- the whole remains stable only because reserve is being consumed?

Ask:

What ratio would be faithful now?

Not permanently.

Not universally.

Now.

Remain with one final question:

**What must I preserve from the previous recursion—and what must I differentiate so that the next return does not become mere repetition?**

## Constitutional Declaration of Recursive and Proportional Integrity

We will distinguish recursion from repetition.

We will distinguish iteration from learning.

We will distinguish recurrence from identity.

We will define every state.

We will define every update.

We will define every parameter.

We will define every initial condition.

We will preserve path dependence.

We will not pretend that transformation order never matters.

We will identify fixed points.

We will test their stability.

We will not call every stable fixed point desirable.

We will identify periodic orbits.

We will distinguish rhythm from immobility.

We will identify bifurcations.

We will not use mathematical transformation as automatic proof of moral or spiritual transformation.

We will distinguish chaos from randomness.

We will distinguish determinism from perfect predictability.

We will estimate sensitivity carefully.

We will preserve the limitations of finite data.

We will distinguish universality from identity.

We will identify the mathematical class within which a universal relation holds.

We will not export one universality class across every domain.

We will define every scale.

We will define every scale map.

We will declare what coarse-graining preserves.

We will declare what coarse-graining destroys.

We will not treat renormalization as a universal metaphor with empirical authority everywhere.

We will preserve the governing rule:

**The pattern may recur. The mechanism changes.**

We will define the pattern.

We will define the mechanism.

We will measure structural resemblance.

We will measure mechanism difference where possible.

We will not let resemblance impersonate causation.

We will distinguish exact self-similarity from statistical self-similarity.

We will distinguish affine similarity from uniform scaling.

We will distinguish hierarchy from fractality.

We will distinguish recursion from fractality.

We will not declare a fractal because an image appears intricate.

We will declare every scaling range.

We will declare every dimension estimator.

We will report uncertainty.

We will test alternative models.

We will not infer infinite scaling from finite observation.

We will preserve the mathematical integrity of iterated function systems.

We will preserve the assumptions required for their fixed sets.

We will not call every natural form an exact iterated-function-system attractor.

We will distinguish similarity dimension from box-counting dimension.

We will distinguish box-counting dimension from Hausdorff dimension.

We will not treat dimension as a universal score of complexity, consciousness, health, or worth.

We will define every ratio.

We will define its orientation.

We will define its units.

We will protect against zero denominators.

We will distinguish composition from magnitude.

We will honor the early inheritance of measure and proportion.

We will practice continuity without anachronism.

We will not assign modern theories to ancient peoples without evidence.

We will recognize the mathematical significance of the golden ratio.

We will preserve its exact definition.

We will not convert it into a universal law of beauty, biology, health, or civilization.

We will not round measurements until they imitate $\varphi$.

We will recognize that unequal proportion may remain coherent.

We will distinguish balance from equality.

We will distinguish equality of dignity from equality of function.

We will not impose fifty-fifty division upon every dynamic system.

We will define what is being balanced.

We will define generation.

We will define loss.

We will declare every system boundary.

We will disclose external support.

We will disclose exported loss.

We will distinguish point balance from cycle balance.

We will distinguish bounded balance from exact equilibrium.

We will distinguish adaptive balance from arbitrary relativism.

We will recognize that a steady state may contain active internal flux.

We will not confuse zero accumulation with zero activity.

We will test the stability of every balance.

We will not call an unstable equilibrium safe.

We will preserve component contributions.

We will not allow an aggregate total to hide local collapse.

We will identify target ratios.

We will justify them.

We will permit them to change when reality changes.

We will not change them merely to excuse failure.

We will preserve inheritance.

We will permit differentiation.

We will not worship repetition.

We will not worship novelty.

We will ask what remains invariant.

We will ask what becomes genuinely new.

We will distinguish latent capacity from emergent capacity.

We will model branching.

We will preserve the uniqueness of descendants.

We will preserve temporal recursion.

We will preserve scale recursion.

We will not confuse time with scale.

We will compare evolve-then-coarse-grain with coarse-grain-then-evolve.

We will measure the commutation defect.

We will not invoke emergence before examining omitted variables and inadequate scale maps.

We will preserve multiscale responsibility.

We will not call one level coherent while it destroys another.

We will preserve the architectural status of holofractal language.

We will not claim exact fractality where only recursive relationship has been shown.

We will preserve the symbolic status of the Spirit Field.

We will not convert interior recursion into a physical fractal field.

We will distinguish what is measured.

We will distinguish what is modeled.

We will distinguish what is inferred.

We will distinguish what is lived.

We will allow every equation to reveal its status.

We will allow every model to reveal its failure conditions.

We will permit reality to correct the pattern we believed we saw.

**The past supplies inheritance.**

**Recursion carries it forward.**

**Differentiation introduces novelty.**

**Ratios make relation measurable.**

**Scale reveals what persists.**

**Fractal mathematics measures specified forms of recurrence.**

**Bifurcation reveals changing possibility.**

**Dynamic balance preserves becoming without requiring equality.**

**The pattern may recur.**

**The mechanism changes.**

**Reality remains the measure.**

## Crossing the Next Arch

Chapter Twenty-Five has now established:

- recursion;
- iteration;
- recurrence relations;
- fixed points;
- periodic orbits;
- bifurcations;
- deterministic chaos;
- Lyapunov exponents;
- period-doubling universality;
- scale transformations;
- renormalization fixed points;
- structural recurrence;
- mechanism difference;
- iterated function systems;
- exact self-similarity;
- statistical self-similarity;
- similarity dimension;
- box-counting dimension;
- finite scaling ranges;
- ratios;
- compositions;
- the golden ratio;

- dimensionless proportionality;
- point balance;
- cycle balance;
- flux balance;
- adaptive balance;
- ratio control;
- recursive differentiation;
- inheritance and novelty;
- dual-aspect recursion;
- scale–dynamics commutation;
- multiscale coherence balance.

Yet self-similarity does not answer the part–whole question.

A fractal part may resemble the whole geometrically.

A distributed representation may preserve information about the whole without resembling it.

A damaged memory system may retain enough distributed encoding for partial reconstruction.

A network fragment may preserve a motif without preserving full organization.

A hologram may permit degraded reconstruction from a partial region.

A code may distribute one message across many symbols.

A culture may preserve civilization across many persons, none of whom contains the complete inheritance alone.

The next formal question therefore becomes:

**How can information about a larger organization be distributed across its parts, and under what conditions can the whole be partially reconstructed after local loss?**

Part III must now distinguish:

- self-similarity from distributed encoding;
- redundancy from holography;
- compression from reconstruction;
- local information from whole containment;
- associative memory from exact storage;
- boundary encoding from literal whole-in-every-part;
- physical holography from architectural analogy.

Recursion has explained how organization may be carried forward.

Chapter Twenty-Six will ask how organization may be carried throughout.

# CATHEDRAL THREAD

Before recursion could become formal, Chapter Twenty-Four had to distinguish state from observation.

Before later organization could inherit earlier organization, a state-transition rule had to become visible.

Before recurrence could become meaningful, exact return had to be distinguished from transformed return.

Before simple rules could be trusted, nonlinear dynamics had to reveal that simplicity of equation does not guarantee simplicity of history.

Before chaos could enter the Cathedral, determinism had to be distinguished from predictability and unpredictability from randomness.

Before cross-scale similarity could become disciplined, universality had to be distinguished from mechanism identity.

Before fractal language could enter SDAE, exact self-similarity had to be distinguished from statistical scaling, and scaling from appearance.

Before recursive differentiation could become a formal architecture, inheritance had to be joined to novelty without reducing either one to the other.

Before ratio could become a teaching device, orientation, units, denominator, and context had to become explicit.

Before the golden ratio could be honored, its exact mathematics had to be separated from universal mystification.

Before balance could become a constitutional principle, equality of dignity had to be distinguished from equality of contribution.

Before dynamic balance could become intelligible, steady state had to be distinguished from inactivity, point balance from cycle balance, and balance from stability.

Before multiscale coherence could be claimed, local success had to be prevented from hiding cross-scale depletion.

The second principle of Part III is now established:

**Organized becoming is recursive when later states inherit and transform earlier organization; it becomes fractal only when a specified form of scaling or self-similarity is mathematically demonstrated.**

The recursive principle follows:

**Recursion carries history forward without requiring the future to repeat the past.**

The nonlinear principle follows:

**Simple deterministic rules can generate fixed points, cycles, bifurcations, and chaos depending upon parameter and initial condition.**

The chaos principle follows:

**Deterministic organization may remain bounded while becoming practically unpredictable across long horizons.**

The universality principle follows:

**Distinct systems may share scaling architecture within a defined class without sharing material mechanism or ontology.**

The scale principle follows:

**Changing scale can preserve selected constraints while discarding detail and introducing new effective variables.**

The Project Cathedral principle follows:

**The pattern may recur. The mechanism changes.**

The fractal principle follows:

**Fractal language must be earned through a declared scaling relation, scale range, dimension, and uncertainty.**

The historical-ratio principle follows:

**Human beings learned to make relation measurable through construction and proportion long before modern symbolic formalism.**

The golden-ratio principle follows:

**The golden ratio is a precise recursive proportion whose unequal parts preserve one self-consistent relation; it is not the required ratio of every coherent system.**

The dynamic-balance principle follows:

**Balance is zero net incoherence in a declared relation—not equal contribution from every differentiated part.**

The rhythmic principle follows:

**A system may preserve itself across a complete cycle even while generation and loss remain unequal at every moment within that cycle.**

The recursive-differentiation principle follows:

**Each stage may inherit selected organization while introducing capacities unavailable to simple repetition of the prior stage.**

The dual-aspect principle follows:

**Exterior and interior histories become recursive when each state alters the conditions through which the next state will be measured, interpreted, and lived.**

The commutation principle follows:

**When evolve-then-coarse-grain differs from coarse-grain-then-evolve, the scale description requires correction, additional variables, or a different effective mechanism.**

The proportionality principle follows:

**Coherence does not require identical parts, identical scales, or identical burdens; it requires relations proportioned sufficiently for differentiated participation to continue.**

The part-whole principle now approaches:

**Recursion explains how the pattern continues through time and scale; distributed encoding must now explain how organization may remain recoverable throughout differentiated parts.**

# RECURSION, SCALE, RATIO, AND DYNAMIC BALANCE



*Recursion carries inheritance forward without requiring exact repetition.*

**RECURSION != REPLICATION | BALANCE != UNIVERSAL 50/50**

FS-G22.01 EQ-25.01

### RELATIONAL UPDATE

Declare the state, update rule, inputs, constraints, and uncertainty before iteration.

FS-G22.02 EQ-25.02

### ITERATED RECURSIVE HISTORY

Preserve order, memory, path dependence, and the distinction between rule and realized trajectory.

FS-G22.03 EQ-25.03

### RECURSIVE DIFFERENTIATION

Track inherited organization, divergence, specialization, novelty, and continuing relation.

FS-G22.04 EQ-25.04

### RECURSIVE SCALE MAP

Declare fine and coarse variables, retained information, lost detail, and reconstruction limits.

FS-G22.05 EQ-25.05

### SCALE COVARIANCE AND POWER LAW

State the finite scaling window, fitted exponent, residuals, alternatives, and predictive test.

FS-G22.06 EQ-25.06

### CROSS-SCALE DUAL-ASPECT CONSISTENCY

Compare scale transformation before and after exterior/interior projection without forcing exact commutation.

FS-G22.07 EQ-25.07

### SELF-SIMILARITY RESIDUAL

Measure the failure of normalized resemblance while keeping mechanism and finite-size limits visible.

FS-G22.08 EQ-25.08

### BOX-COUNTING DIMENSION

Declare estimator, segmentation, scale range, threshold sensitivity, and uncertainty.

FS-G22.09 EQ-25.09

### TARGET RATIO AND LOG-RATIO ERROR

Name numerator, denominator, orientation, units, target authority, horizon, and noncompensable limits.

FS-G22.10 EQ-25.10

### THE MA'AT RATIO

Use Ma'at as ancestral-symbolic right relation, never as a physical field, hidden mechanism, or ancient modern equation.

FS-G22.11 EQ-25.11

### RATIO DYNAMICS

Track moving targets, relative rates, phase, delay, overshoot, and scale-dependent consequence.

FS-G22.12 EQ-25.12

### STRUCTURAL BALANCE VERSUS RATE EQUALITY

Distinguish unequal viable proportion from equality of commensurate opposing rates under a declared stationarity condition.

**DIGNITY | UNCERTAINTY | SOURCE TRANSPARENCY | FAILURE | REVISION**

**Cross-reference equations: EQ-25.13, EQ-25.14 | FS identifies figure position; EQ identifies canonical equation. Reality remains the measure.**

www.ingramcontent.com/pod-product-compliance
Lightning Source LLC
LaVergne TN
LVHW051920051126
851857LV00021B/2089
*9798952413016*